Operational amplifiers

I. E. SHEPHERD

C.Eng., M.I.E.R.E., M. Inst.M.C.

Senior Scientific Officer, Hydraulics Research Station,
Wallingford, Oxfordshire, U.K.

LONGMAN

LONDON AND NEW YORK

Longman Group Limited,
Longman House,
Burnt Mill, Harlow, Essex

*Published in the United States of America
by Longman Inc., New York*

© Longman Group Limited 1981

First published 1981

British Library Cataloguing in Publication Data
Shepherd, I E
Operational amplifiers.
1. Operational amplifiers
2. Linear integrated circuits
I. Title
621.381'73'5   TK7871.58.06   80–40770

ISBN 0–582–46089–1

Typeset by Macmillan India Limited,
Bangalore

Printed in Singapore by Kyodo Shing Loong Printing Industries Pte. Ltd.

Contents

List of Tables

Preface

The ability to design linear and non-linear circuits using operational amplifiers continues to be an important skill in spite of the enormous advances in digital techniques and microprocessing devices. Many subtle pitfalls lurk for the unwary and inexperienced designer, and many are misled by the apparent simplicity of analog circuits. A thorough knowledge of the limitations as well as the advantages of the many devices now available is necessary for the realisation of predictable designs and the choice of the best chip for the job.

Design experience affords a judgement about what is or is not important; the uninitiated often find that a design does not perform as expected due to lack of consideration of a vital parameter. In this book I have set out many of my own personal experiences and lessons learned the hard way – by making mistakes. I hope you will benefit from them. I have often found the literature inadequate when trying to understand mathematical proofs when solutions are assumed. I have therefore tried to include as much working as possible, with additional proofs in the Appendix. The mathematics is definitely not elegant, and for that I apologise. The methods will be disputed and unacceptable to some. To those readers I can only explain that the solutions have been my own way of producing the results to my own satisfaction. It has not been sufficient for me to accept the traditional methods without a personal approach, and therefore any mistakes and errors of concept are entirely mine.

Manufacturer's data and applications information are invaluable sources of guidance. New devices are continually emerging, and a current awareness is vital. I have primarily considered single chip devices, but many hybrid, multi-chip packages will be found to offer advantages. There are, of course, many applications not included but I hope that those discussed will be of benefit.

To all my past and present colleagues I say 'thank you' for the willing advice and assistance they have given; and to Joe Collins the credit for determining my attitude and approach to circuit design generally.

I.E.S. 1981

Acknowledgements

We are grateful to the following for data reproduced in this book: Mullard Ltd; Precision Monolithics Inc.; National Semiconductor (UK) Ltd; Analog Devices Ltd; Burr-Brown International Ltd; Raytheon Company; RCA Ltd; Rastra Electronics Ltd; Texas Instruments Ltd; Teledyne Philbrick; Fairchild Camera and Instrument (UK) Ltd; Intersil Datel (UK) Ltd; Motorola Ltd; and Harris Semiconductor.

1

Operational amplifiers – the broader view

1.1 Background

There have been many milestones in the evolution of electronic engineering. The necessity to achieve a major objective has often been the spur to invention. Examples include telephony, telecommunications, radio and television, military and space research – all of which have demanded rapid and extensive changes in the type and performance of electronic components. The human propensity for innovation has produced the methods, sometimes by intention, sometimes by accident, but frequently contemporaneously with requirements. Dummer has given an interesting review of the evolution of electronic components.[1] Many names are familiar to the engineer because they are used in electronic terminology – Volta, Ampère, Ohm, Faraday, Henry – and others are household names – Edison, Marconi, Baird, and so on. Men who have most recently made major contributions to the evolution of electronics as we know it today are perhaps less familiar to the non-specialist – Bardeen, Brattain and Shockley (transistor), Hall (Hall effect), Hoeni (planar process), Kilby (integrated circuit) and Widlar (linear IC amplifiers) are a few.

Arguably one of the most important events in the evolution of electronics was the introduction of the telegraph. In 1837 Cooke and Wheatstone were granted British Patent No. 7390 for a 'Telegraph Bell Relay', and in 1838 Edward Davy's 'Telegraph Signal Relay' was granted British Patent No. 7719.

Other inventions followed, and in 1904 Fleming introduced the two-electrode vacuum tube. This device, succeeded by the three-electrode tube invented by De Forest in 1906, made possible the whole field of voltage regulation and signal amplification which were the fundamentals of electronics. Much of the early development of amplifiers, and in particular feedback techniques, was aimed at improving telephone communications.

1

The present diversification of operational amplifier applications with which we are now familiar can be traced to the early linear integrated circuit amplifiers such as the Fairchild μA702 or Texas SN521, although analog computers were, of course, constructed with valves in the early days. Between these periods, discrete transistor amplifiers were extensively used, but their life in historical terms would have been relatively short, since transistors were not readily available until about 1950 and the first linear integrated circuits appeared in about 1962. Some idea of the improvement in linear ICs can be seen in Table 1.1.

Today we have available a very wide choice of differing operational amplifiers (abbreviated throughout this book as 'op amps'), and to the inexperienced engineer it is a virtually impossible task to choose the device most suited to a given application. Even if he is capable of making a selection on technical merit, there are many hidden problems waiting to trap the unwary. Not the least of these is interpretation of specifications, closely followed by such diverse considerations as cooling, package configuration, availability, price, estimation of spreads where typical values are quoted, reliability in adverse environments, radiation susceptibility, etc. In the words of a recent report 'underspecifying can mean disaster, overspecifying costs money'.[2] It is well worth while spending some time attempting to find the best compromise, especially in a production environment where the profitability of an end product depends on optimisation of parts costing. I hope that this book will demonstrate some of the limitations of op amps in various common applications and thereby assist in this elimination process. Some of the less obvious considerations are discussed in more detail below.

1.2 Current awareness

Linear op amps are under continuous development, and improvements of performance are frequently announced. It is, therefore, imperative that the practising circuit-design engineer should read the technical literature as a matter of course. This point is emphasised by Table 1.1, which is a very small selection of some of the principal stages of op amp evolution showing the steady progress towards the 'ideal' amplifier. There is no reason to suppose that this is the end of the line.

1.3 Second sourcing

Anyone involved with electronic equipment manufacture will be aware of the problems created if it is found that the single supplier of

Table 1.1 *Some important advances in operational amplifiers.*

Device type	Approx. date of introduction	Operable supply voltage range	Max input offset voltage (mV)	Internal compensation	Typical input Voltage range at nominal supply	Typical open-loop gain at nominal supply (V/V)	Max input bias current (nA)	Max power supply at nominal voltage (mW)	Min CMRR (dB)
μA702C	1963	$\begin{Bmatrix}-3-6\\+6+12\end{Bmatrix}$	6	×	$-4+0.5$*	3.4×10^3	7 500	125	70
NE515	1965	7 to 12 Total	3	×	$-1+1.5$§	3.2×10^3	31 000	50	100 ‖
μA709C	1965	$\pm9\pm15$	7.5	×	±10†	4.5×10^4	1 500	200	65
LM201	1967	$\pm5\pm20$	7.5	×	±12 min†	1.5×10^5	1 500	90	65
μA741C	1968	$\pm2\pm15$	6	✓	±13†	10^5	500	85	70
μA727B	1968	$\pm9\pm15$	10	×	±13†	10^2	12	330	100
LM308	1969	$\pm2\pm15$	7.5	✓	±14 min†	3×10^5	7	24	80
UC4250C	1969	$\pm1\pm18$	6	✓	±13.5† min	6×10^4 min	80	3 ¶	70
μA725	1969	$\pm3\pm22$	2.5	×	±14†	3×10^6	125	150	115 typ
LM318	1970	$\pm5\pm20$	10	✓	±11.5 min	2×10^5	500	300	70
AD504J	1972	$\pm5\pm18$	2.5	×	$\pm Vs$†	4×10^6	200	120	94
LF355	1976	$\pm5\pm20$	10	✓	$-12+15.1$†	2×10^5	0.2	120	80
CA3160	1976	$\pm2.5\pm8$	15	✓	$-8+4.5$‡	3.2×10^5	0.05	225	70
TL061C	1976	$\pm1.5\pm18$	15	✓	±10 min†	10^4	0.4	7.5	70
OP-15G	1977	$\pm5\pm20$	3	✓	$-11.5+14.8$†	2×10^5	0.4	150	82
AD517	1978	$\pm5\pm18$	0.15	✓	–	10^6 min	5	120	94
LM10	1979	1.1 to 40	4	✓	$-15+14.15$†	2.5×10^5	30	15	90
ICL7600	1979	$\pm2\pm8$	0.005	✓	±6.5‡	10^5	3	50	88

All commercial device specifications

* $Vs = +12-6$
† $Vs = \pm15$
‡ $Vs = \pm7.5$
§ $Vs = -3+6$
‖ Differential output.
¶ Can be operated at 24 μW. Bias current = 10 nA.

an op amp selected for a design cannot deliver at the rate required. It is desirable to establish the prospect of continuing availability from an alternative source of supply whenever possible.

1.4 Pin-for-pin compatibility

Some devices are advertised as being pin-compatible with other standard types. This can often be an advantage since it is possible to update and improve the performance of existing circuits by replacing op amps with new types.

1.5 Special acceptability tests

Many applications demand increased environmental durability and reliability. Many manufacturers can carry out in-house testing to military specifications (e.g. MIL, DEF) and to requirements under the BS9000 scheme. Devices can be given qualification approval and released to these standards by the manufacturer, and usually a price increase is applicable. Device failure-rate and quality-acceptance tests can be specified. Many data-sheet parameters are not 100 per cent tested, so if guaranteed performance and reliability are required it is important to check with the manufacturer.

There are often two and sometimes three operating temperature ranges over which the same basic device is specified (e.g. 0 to $+70\,°C$, -25 to $+85\,°C$, -55 to $+125\,°C$). Sometimes the same basic performance is applicable over each range, but for others degraded parameters are specified over smaller ranges. For example, improved performance may well be achieved by selecting a -55 to $+125\,°C$ device even if the actual operating temperature range is only $10-50\,°C$.

Packaging is another specialised consideration. Although many standard mechanical outline configurations are established, the actual method of encapsulation varies enormously, with many manufacturers having their own proprietary processes. Again, failure rates and testing should be established where high reliability is important.

1.6 Cost effectiveness

For many applications there is a wide choice of amplifier types which will provide adequate accuracy. The choice then becomes a matter of economics. Buying from distributors is often the best method since a ready source of short delivery is available, which helps to reduce the cost of stocking large quantities of components.

It is the buyer's function to obtain parts at lowest cost whilst at the same time establishing continuity of delivery. It is the designer who should consider the technical alternatives for cost minimisation. For example, if four amplifiers are required, is a quad package cheaper per channel than single devices? Use of amplifiers in existing designs could also influence choice.

1.7 Specmanship

This is the term used to describe the technique by which products are specified to emphasise their advantages, but where limitations of performance are either not stated or are difficult to interpret. I am not suggesting that any manufacturers intentionally mislead; rather that data is presented in its best light and that great care is often required to make an adequate interpretation of the information provided.

Let us take an example. FET input amplifiers have very low bias currents, which are, in fact, leakage currents. The bias current is temperature-dependent, and is often quoted at 25 °C. However, when the amplifier is operating, its own internal dissipation heats the chip above ambient, to a level dependent on the case thermal resistance and any heat sinking present. The bias current for a chip temperature of 25 °C can be significantly different from the bias current at an ambient temperature of 25 °C. This demonstrates the care with which data sheets must be treated.

It is often found that typical values are quoted rather than maxima and minima, and this renders critical design parameters difficult to achieve on a production basis. Even where limits are specified, the distribution within the limits is not known, so the user will need to resort to his own in-house testing facility.

1.8 Alternative analysis techniques

There are at least three methods which can be used to analyse an op amp circuit, and throughout the text I have used the method which either helps to reduce the complexity or produces a flexible approach to a specific circuit problem. I do not claim that the methods shown are the only ways to solve the problems. Three of the most used techniques are shown below.

1.8.1 Virtual earth, current equating

In Fig. 1.1 it is assumed that the amplifier input current is zero and the input resistance infinite. If $V_1 = 0$, i.e., a 'virtual earth' point (the

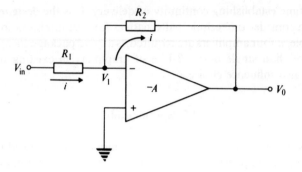

Fig. 1.1 Basic op amp configuration

magnitude of V_1 is a function of A, and only approaches zero as A approaches infinity), then

$$i = \frac{V_{in}}{R_1} = \frac{-V_o}{R_2}$$

$$\therefore \frac{V_o}{V_{in}} = \frac{-R_2}{R_1}$$

This approach is often most convenient when multiple input connections are used.

1.8.2 Equating input voltages

Assume that the amplifier input resistance is infinite and that source resistance and output resistances are zero, then

$$V_1 = \frac{R_2 \, V_{in}}{R_1 + R_2} + \frac{R_1 \, V_o}{R_1 + R_2}$$

If A is infinite, the voltage difference between the inverting and non-inverting amplifier inputs is zero.

$$\therefore V_1 = 0 \text{ and } \frac{R_2 \, V_{in}}{R_1 + R_2} = \frac{-R_1 \, V_o}{R_1 + R_2}$$

$$\therefore \frac{V_o}{V_{in}} = \frac{-R_2}{R_1}$$

This method of analysis is often useful for non-linear and oscillator circuits.

1.8.3 Feedback amplifier gain

If the output resistance is zero,

$$V_1 = \frac{R_2 V_{in}}{R_1 + R_2}$$

V_1 is the effective input voltage to an amplifier which then has negative feedback applied to it. Feedback fraction $\beta = R_1/(R_1 + R_2)$ if the source resistance is zero. Closed-loop gain of an amplifier with negative feedback is

$$G = \frac{-A}{1 + \beta A} \approx \frac{-1}{\beta} \text{ if } A \gg 1$$

$$\therefore V_o = \frac{-1}{\beta} \cdot V_1 = \frac{-R_2}{R_1} V_{in}$$

Note that using this method implies that V_1 is the effective voltage, not the actual voltage present at the input terminal.

The assumptions made above can often be justified since they involve small errors in the result. However, in order to demonstrate the typical magnitude of the errors, later chapters examine in depth the effects of finite open-loop gain, input resistance and non-zero output resistance. It is only by considering these errors that the engineer can assess the approximations made. In practice the limiting factors often depend on the lack of precision of external components, since it appears that the performance of new amplifiers has advanced faster than that of discrete components.

2

Ideal operational amplifiers and practical limitations

2.1 Definitions

In order to discuss the ideal parameters of operational amplifiers, we must first define the terms, and then go on to describe what we regard as the ideal values for those terms. At first sight, the specification sheet for an operational amplifier seems to list a large number of values, some in strange units, some interrelated, and often confusing to those unfamiliar with the subject. The approach to such a situation is to be methodical, and take the necessary time to read and understand each definition in the order that it is listed. Without a real appreciation of what each means, the designer is doomed to failure. The objective is to be able to design a circuit from the basis of the published data, and *know* that it will function as predicted when the prototype is constructed. It is all too easy with linear circuits, which appear relatively simple when compared with today's complex logic arrangements, to ignore detailed performance parameters which can drastically reduce the expected performance.

Let us take a very simple but striking example. Consider a requirement for an amplifier having a voltage gain of 10 at 50 KHz driving into a 10 KΩ load. A common low-cost, internally frequency-compensated op amp is chosen; it has the required bandwidth at a closed-loop gain of 10, and it would seem to meet the bill. The device is connected, and it is found to have the correct gain, but it will only produce a few volts output swing when the data clearly shows that the output should be capable of driving to within two or three volts of the supply rails. The designer has forgotten that the maximum output voltage swing is severely limited by frequency, and that the maximum low-frequency output swing becomes limited at about 10 KHz. Of course, the information is in fact on the data sheet, but its relevance has not been appreciated. This sort of problem occurs regularly for the inexperienced designer. So the moral is clear: always take the

8

necessary time to write down the full operating requirements before attempting a design. Attention to the detail of the performance specification will always be beneficial. It is suggested the following list of performance details be considered:

1 Closed-loop gain accuracy, stability with temperature, time and supply voltage.
2 Power supply requirements, source and load impedances, power dissipation.
3 Input error voltages and bias currents. Input resistance, output resistance. Drift with time and temperature.
4 Frequency response, phase shift, output swing, transient response, slew rate, frequency stability, capacitive load driving, overload recovery.
5 Linearity, distortion and noise.
6 Input, output or supply protection required. Input voltage range, common-mode rejection.
7 External offset trimming requirement.

Not all of these terms will be relevant, but it is useful to remember that it is better to consider them initially rather than to be forced into retrospective modifications.

ALL PARAMETERS ARE SUBJECT TO WIDE VARIATIONS

Never forget this fact. How many times has a circuit been designed using typical values, only to find that the circuit does not work because the device used is not typical? The above statement thus poses a tricky question: when should typical values and when should worst-case values be used in the design? This is where the judgement of the experienced designer must be brought to bear. Clearly, if certain performance requirements are mandatory, then worst-case values must be used. In many cases, however, the desirability of a certain defined performance will be a compromise between ease of implementation, degree of importance, and economic considerations.

DO NOT OVER-SPECIFY OR OVER-DESIGN

In the end, we are all controlled by cost, and it is really pointless taking a sledgehammer to crack a nut. Simplicity is of the essence, since the low parts count implementation is invariably cheaper and more reliable.

As an example of this judgement about worst-case design, consider a low-gain d.c. transducer amplifier required to amplify 10 mV from a voltage source to produce an output of 1 V with an accuracy of ± 1 per cent over a temperature range of $0-70\,°C$. Notice that the specification calls for an accuracy of $\pm$ 1 per cent. This implies that the output should be 1 V $\pm$ 10 mV from 0–70 °C. The first step is, of course, to consider our list above, and decide which of the many parameters are relevant. Two of the most important to this (very limited) specification are offset voltage drift and gain stability with temperature. We will assume that all initial errors are negligible (rarely the case in practice). The experienced designer would know that most op amps have a very large open-loop gain, usually very much greater than 10 000. A closed-loop gain change of $\pm$ 1 per cent implies that the loop gain (as explained later) should change by less than $\pm$ 100 per cent for a closed-loop gain of 100. This is clearly so easily fulfilled that the designer knows immediately that he can use typical open-loop gain values in his calculations. However, offset voltage drift is another matter. Many op amp specifications include only typical values for offset voltage drift; this may well be in the order of 5 μV/°C, with an unquoted maximum for any device of 30 μV/°C. If by chance we use a device which has this worst-case drift, then the amplifier error could be $30 \times 70 = 2.1$ mV over temperature, which is a significant proportion of our total allowable error *from all sources*. (**Beware**: the nature of these parameters is that they are not linearly related to temperature, and so the method of specification must be considered, e.g. average slope, maximum slope, etc.)

Here is a case, then, where one can be confident that the typical value of open-loop gain can be used, but where the maximum value of drift may well cause significant errors. This sort of judgement is essential in careful design, and great care is required in interpreting manufacturers' data. This consideration must be extended to all the details listed above; apart from the fact that worst-case values are often not quoted, it is often found that values given are not 100 per cent tested. Statistical testing is employed which, for example, guarantees that 90 per cent of all devices fall within the range specified. It could be very inconvenient for the user who relies on the specified performance and then finds that he has several of the 'other' 10 per cent actually plugged into his circuit.

Let us consider now the fundamental op amp definitions. This list includes only the 'operating' parameters, not the 'without damage' maximum ratings which are discussed elsewhere.

2.1.1 *Open-loop gain* (large-signal voltage gain, differential voltage gain)

This is the ratio of the amplifier output voltage change to the voltage change between its input terminals when no feedback is applied, and is normally specified at low frequency under defined power supply, temperature and loading conditions. The op amp chip itself may well have internal feedback, but the user does not need to know this.

2.1.2 *Input voltage range* (maximum common-mode voltage)

This is the voltage range on either input terminal within which the amplifier will continue to function correctly. It is not necessarily the same parameter as the maximum differential input voltage. For some single-supply type op amps operating between 0 V and $+V_{cc}$ (the positive supply), the input range actually extends down to 0 V since the input transistors operate with common collectors. Some FET input amplifiers can operate with the input terminals below the supply low.

2.1.3 *Common-mode rejection ratio* (common-mode gain)

An ideal differential amplifier will produce zero output voltage change when both input terminals are subjected to the same source voltage change. Since each input of a practical amplifier does not produce exactly the same internal gain, when they are differenced an error voltage remains which appears at the output terminal. A measurement can be made of the output error voltage for a commonly applied input, and the result is a means of showing how near to ideal the differential function operates. When the output error is divided by the amplifier gain, we obtain an equivalent input error voltage produced by the common mode (CM) voltage. The common-mode rejection range (CMRR) is then the ratio in dB of the common-mode input to the input-referred error voltage.

For example, if CM input is $+1$ V, output change is $+0.1$ V, and amplifier gain is 100, the input-referred error is $0.1/100 = 1$ mV. Then the CMRR $= 1$ V$/1$ mV $= 1\,000 = 60$ dB.

2.1.4 *Input impedance*

Refer to Fig. 2.1, which shows the input circuit. Both the differential and common-mode input connections have a resistive and capacitive part to their values and can be listed separately.

2.1.4.1 Differential input resistance (input resistance). The meaning of this is not consistent between manufacturers. In some cases it means the resistance measured between the two input terminals, and in others the input resistance to one terminal with the other grounded.

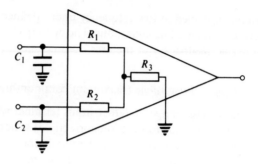

Differential input resistance	$= R_1 + R_2$	(Between inputs)
	$= R_1 + \dfrac{R_2 R_3}{R_2 + R_3}$	(one input grounded)
Common-mode input resistance	$= R_3 + \dfrac{R_1 R_2}{R_1 + R_2}$	(both inputs connected together)
Differential input capacitance	$= \dfrac{C_1 C_2}{C_1 + C_2}$	
Common-mode input capacitance	$= C_1 + C_2$	

Fig. 2.1 Op amp input circuit

In most cases the former will be approximately twice the latter. In each case it is usually measured at 25 °C, with the input stage balanced and at zero common-mode voltage.

2.1.4.2 Differential input impedance. This is similar to the above, but either at a defined frequency or a specified value of input capacitance.

2.1.4.3 Common-mode input resistance. Again, as in Section 2.1.4.1 above, there are some differences of definition. In some cases it is the input resistance with both input terminals connected together, and in others between either input terminal and ground.

2.1.4.4 Common-mode input impedance. This is the effect of CM input capacitance as in Section 2.1.4.2. Common-mode input parameters are non-linear functions of common-mode voltage and temperature, so all values are quoted under defined conditions. Common-mode input resistance of FET input amplifiers reduces by a factor of 2 for every 10 °C temperature rise. Incremental values may be significantly different from the large-signal average values which are usually quoted.

2.1.5 Output resistance

This is the resistance measured 'looking back' into the output terminal under small-signal, low-frequency, null output conditions. If the output resistance is significant compared with the load, the effective open-loop gain may be reduced.

2.1.6 Input offset voltage

When both input terminals are shorted together, there will be an output voltage present. This is due to the inherent mismatch between the devices connected to each input terminal. To reduce this output to zero, it is necessary to apply a small differential input signal, and the value of this is the input offset voltage.

2.1.7 Input offset current (initial difference current)

The difference between the input bias currents flowing into each input terminal, when the output is zero.

2.1.8 Input bias current

This is defined as the value of the current flowing into each input terminal under defined conditions. It is sometimes expressed as the average value of the currents in each input. If offset current (section 2.1.7) is very small, then the bias currents to each input are substantially equal. Input bias currents for bipolar input op amps generally decrease with temperature (as does offset current), but with FET input amplifiers the input current, which is really gate leakage, tends to double for every $10°C$ rise in temperature. Note that if average values are quoted, the actual bias current to any one input could be considerably larger than the value given. Sometimes internal bias current compensation has been applied, in which case the residual bias currents may flow in either direction in the input terminals.

2.1.9 Output voltage swing (full-power response)

There are several points to be considered here. The output swing obtainable depends on supply voltage, frequency, load, and tempera-ture. Typically, with a complementary or quasi-complementary class B type output stage, the output terminal will swing to within two or three volts of each supply rail. This would be at low frequencies and with a defined load value. There is generally output short-circuit protection on the chip which tends to limit the output swing available, and the preceeding driving stage capability will be instrumental in determining the output swing. It is often found that the maximum output voltage swing is only obtainable at low currents, whereas considerably more current can be driven at lower voltage swings. For example, with a 15 V supply, the output capability may be 12 V into 10 KΩ (i.e. 1.2 mA) but 10 V into 2 KΩ (i.e. 5 mA). This non-linear effect is due to the saturation of the output transistor.

Output swing capability also depends on frequency, and tends to be related to 'slew rate', which is defined later (section 2.1.12). Again we assume a purely resistive load (capacitive loading is another matter). This limitation is often due to internal capacitors (possibly frequency-compensation components) and the capability of the active devices to drive increasing currents into the capacitors as the frequency goes up. The output swing at increasing frequencies will also depend on the values of external components selected for frequency-compensation purposes. Hence if large signal swings are required at anything above a few kilohertz, attention must be paid to this limitation. Do not confuse 'bandwidth' (meaning small-signal bandwidth) with large-voltage swing capability. The small-signal bandwidth may be many times greater than the large-signal bandwidth.

Some manufacturers use the term 'full-power response', which is subdivided into full linear response and full peak response. The former is based on a maximum swing for which distortion does not exceed an arbitrary fixed level, into a defined load. The latter is based on the maximum frequency at which the full output swing can be expected, irrespective of distortion or linearity. In some high-frequency applications, linear waveforms are not of fundamental importance, so this latter parameter is more useful. Note that crossover distortion increases as a percentage of the output level for lower output swings, especially at high closed-loop gains. Where the loop gain is fairly high, crossover distortion is minimised by the effects of feedback and can often be ignored. Linearity is often important in transducer amplifier applications. Sometimes an error offset voltage can be generated by rectification of the asymmetrical feedback waveform or overloading the input stage with large signals at the summing junction.

2.1.10 Power supply rejection ratio

When the power supply voltages change, either due to regulation or drift errors, an apparent output signal error is produced. This error can be referred to the amplifier input terminals as an offset voltage. The ratio of supply-voltage change to offset error produced is expressed in dB. Most amplifiers have a high rejection ratio which enables them to be used with poorly regulated supplies at low frequencies, but the rejection tends to decrease with frequency, so that at 100 Hz (typical power-supply ripple) the rejection is not as good as at d.c. It should be noted that where dual supply rails are used the error may be different for one supply compared with the other. This problem is often minimised by using tracking voltage regulators; such regulators also have the great advantage that the input common-mode voltage remains constant with respect to the source.

2.1.11 Bandwidth

Bandwidth is traditionally known as the frequency at which the small signal response decreases to $3\,dB\,(1/\sqrt{2})$ of the nominal low-frequency response. However, this value is only approximate for internally compensated devices operating open-loop and is often determined by the user if external frequency-compensating components are used. Figure 2.2 indicates the differences. The internal capacitors used in op amps tend to have a significant temperature

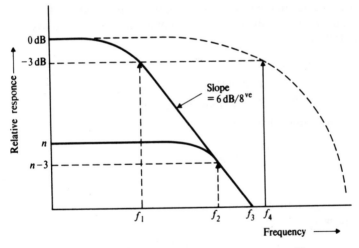

$f_1 = -3\,dB$ point of internally compensated amplifier operating open loop

$f_2 = -3\,dB$ point of internally compensated amplifier with a closed loop gain of $n\,dB$

$f_3 =$ Unity gain frequency of internally compensated amplifier

$f_4 = -3\,dB$ point of uncompensated open loop amplifier

Fig. 2.2 Amplifier bandwidth definitions

dependence, and so the closed-loop bandwidth in such cases is also temperature-dependent.

The gain bandwidth product f_T is a commonly specified parameter for transistors, but a more frequently defined value for op amps is the unity gain frequency f_3. It is a relatively easy parameter to measure and serves as a figure of merit for the comparison of amplifiers. In many cases frequency response is not specified numerically, but all the information is expressed graphically.

2.1.12 Slew rate

In many applications, bandwidth or output voltage swings do not adequately describe the required performance. In analog-to-digital converters, for example, the output may be required to 'slew' from one output limit to another in the minimum possible time, and to drive capacitive loads. In voltage-follower applications, the ability of the output to follow rapidly changing non-sinusoidal input voltages is important. Slew rate is defined as the maximum rate of change of output voltage, and is often given with a specified capacitive load. When the maximum rate of change at the output is *not* limited by the output driving capability, an approximate value can be obtained from the following.

For a sinusoidal output,

$$V_{out} = V_{max} \sin \omega t \qquad \omega = 2\pi f$$

$$f = \text{frequency component of interest}$$

Rate of change of output,

$$\frac{dV_{out}}{dt} = \omega V_{max} \cos \omega t$$

Maximum rate of change of output $= \omega V_{max}$ ... [1]

For a single time-constant equivalent R–C circuit, where $f_0 = -3\,\text{dB}$ frequency,

$$V = \frac{V_p}{\sqrt{1 + (f/f_0)^2}} \qquad\qquad \text{... [2]}$$

V_p = low-frequency peak output voltage
from [1] and [2], maximum rate of change

$$R = \frac{\omega V_p}{\sqrt{1 + (f/f_0)^2}} = \frac{2\pi f V_p}{\sqrt{1 + (f/f_0)^2}} \qquad \text{... [3]}$$

Alternatively, if rise time is defined as $t_r = 2.2CR$ (section 2.1.13) and

$$f_0 = \frac{1}{2\pi CR} \quad \therefore f_0 = \frac{0.35}{t_r}$$

$$R = \frac{2\pi f V_p}{\sqrt{1 + (t_r f/0.35)^2}} \qquad \ldots [4]$$

In practice, of course, the roll-off in response is often not limited to a single time-constant characteristic, so the above can only be a guide towards what can be expected.

2.1.13 Rise time

The rise time is defined as the time taken for the output voltage to rise from 10 to 90 per cent of its total deviation in response to a non-saturating input step function. This does not apply to the overdrive situation. For the sake of completeness, the effective rise time for a single R–C and double (buffered) R–C is derived below.
From Fig. 2.3:

$$V_2 = V_1 \left(1 - e^{-t/C_1 R_1} \right) \qquad \ldots [5]$$

$$V_4 = V_3 \left(1 - e^{-t/C_2 R_2} \right) \qquad \ldots [6]$$

$$\therefore V_4 = V_1 \left(1 - e^{-t/C_1 R_1} \right)\left(1 - e^{-t/C_2 R_2} \right) \qquad \ldots [7]$$

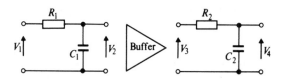

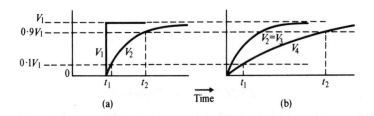

Fig. 2.3 Rise time of R–C networks

For the single R–C, Fig. 2.3(a):

at t_1, $$\frac{V_2}{V_1} = 0.1 = 1 - e^{-t_1/C_1 R_1} \qquad \ldots [8]$$

at t_2, $$\frac{V_2}{V_1} = 0.9 = 1 - e^{-t_2/C_1 R_1} \qquad \ldots [9]$$

From [8], $e^{t_1/C_1 R_1} = 1.111$, and from [9], $e^{t_2/C_1 R_1} = 10$

$$\therefore \frac{t_1}{C_1 R_1} = \log_e 1.111 \quad \text{and} \quad \frac{t_2}{C_1 R_1} = \log_e 10$$

$$\therefore t_2 - t_1 = t_r = C_1 R_1 [\log_e 10 - \log_e 1.111]$$

$$\therefore t_r = 2.2 C_1 R_1 \qquad \ldots [10]$$

For the two R–C cascade circuit of Fig. 2.3(b), a solution has been given by Millman and Taub.[3] It is shown that

$$\frac{V_4}{V_1} = 1 + \frac{1}{T_2 - T_1} (T_1 e^{-t/T_1} - T_2 e^{-t/T_2}) \qquad \ldots [11]$$

where $T_1 = C_1 R_1$ and $T_2 = C_2 R_2$. The solution where $T_1 = T_2 = T$ is unique and gives

$$\frac{V_4}{V_1} = 1 - \left(1 + \frac{t}{T}\right) e^{-t/T} \qquad \ldots [12]$$

To obtain the value of t/T for which $V_4/V_1 = 0.1$ and 0.9, the results are plotted graphically, and for the case where $T_1 = T_2 = T$, $t_1 = 0.55T$ when $V_4/V_1 = 0.1$, and $t_2 = 3.85T$. The rise time is therefore $t_2 - t_1 = 3.3T$ compared with $2.2T$ for the single R–C circuit [10]. The rise time t_r of two identical cascaded (buffered) sections is 1.5 times larger than that of the single circuit. It is shown in Appendix 1 that the $-3\,\mathrm{dB}$ frequency for a double R–C is

$$\omega_1{}^2 = \sqrt{\frac{(T_1{}^2 + T_2{}^2)^2 + 4T_1{}^2 T_2{}^2}{4T_1{}^4 T_2{}^4}} - \frac{(T_1{}^2 + T_2{}^2)}{2T_1{}^2 T_2{}^2} \qquad \ldots [13]$$

If $T_1 = T_2$, $$f_1 = \frac{0.644}{2\pi T} \qquad \ldots [14]$$

$$\therefore t_r = \frac{0.34}{f_1} \qquad \ldots [15]$$

2.1.14 Transient response

This is the closed-loop step function response of the amplifier under small-signal conditions. Notice that this specification applies par-

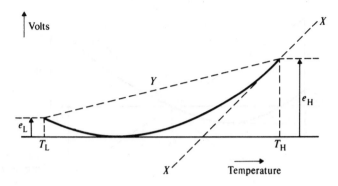

Fig. 2.4 Drift specification

ticularly to the closed-loop situation, whereas that is not so with rise time and slew rate. The transient response will depend on the open-loop frequency response, the effects of loading, stray capacity and phase response. For those who are not familiar with the effects and meaning of phase linearity, the reader is referred to Chapter 3.

2.1.15 Temperature drift
(offset voltage, current, and bias current drift)

These parameters have been considered here because they are all temperature-dependent. Offset bias current tends to decrease with temperature in bipolar amplifiers, but doubles about every 10°C for FET input amplifiers. Offset voltage drift can be of either polarity, and tends to be larger when the initial offset voltage is larger. Beware of the method of specification for drift effects. Generally they are defined as the average slope over a specified temperature range. Figure 2.4 shows one possible method for defining drift.

The solid line represents the true value of the offset voltage with temperature T.

$$\text{Drift} = \frac{e_H - e_L}{T_H - T_L} \text{ volts/°C}$$

This gives the slope shown by the dotted line Y, but is obviously nowhere near the maximum slope of voltage change with tempera-ture, X–X. A more realistic approach is sometimes adopted as shown in Fig. 2.5.

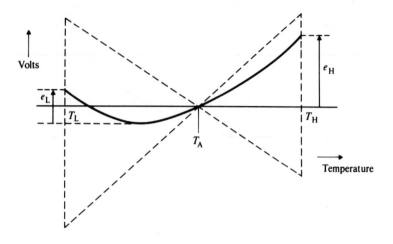

Fig. 2.5 Alternative method for drift specification

With this method, the specified maximum drift value would be either:

(a) $\dfrac{e_H}{T_H - T_A}$ or $\dfrac{e_L}{T_A - T_L}$; or

(b) $\dfrac{|e_H| + |e_L|}{T_H - T_L}$

In this way the total offset computed from temperature range × drift value specified always exceeds the value obtained in practice. The diagonals represent an envelope within which the total offset would remain (assuming initial offset is zeroed at T_A), if the slopes of the diagonals represent the maximum slope of the actual plot.

It can be seen that the method of specifying drift is very important, and when making comparisons between amplifiers of differing manufacture attention must be paid to this fact.

2.1.16 *Settling time* (setting time)

This is fundamentally a closed-loop parameter, and depends on many variables simultaneously. It involves both linear and non-linear components, and depends on a combination of bandwidth, transient response, rise time and slew rate. It is therefore very important to understand the precise meaning of a manufacturer's definition. It can usually be inferred from the following: the time between the initiation of a perfect input step function and the time when the output voltage

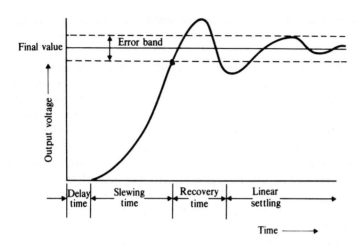

Fig. 2.6 Output settling time

has settled to within a specified symmetrical error band about the final value. Figure 2.6 gives an indication of the effect.

2.1.17 Noise

This is a complex parameter and cannot be defined simply. It therefore merits a more complete discussion later. The equivalent input-referred noise circuit for an op amp consists of current generators, voltage generators and Johnson (thermal) noise. Generally one or the other noise sources can be minimised, but rarely all simultaneously; the end result must always be a compromise. In addition there is the so called 'popcorn' noise, a very important consideration in low-level circuits, of which more later. Popcorn noise is an effect rarely discussed in the literature, and yet in practice it can be more significant than any of the other previously mentioned noise sources. No design would be rigorous without a consideration of power supply, common line, electromagnetic and electrostatically generated and cable noise – although these latter are usually generated externally to the device in question. Some commonly used definitions are:

2.1.17.1 Input noise voltage (current). The peak-to-peak noise voltage (current) in a specified frequency band.

2.1.17.2 Input noise voltage (current) density. The rms noise voltage (current) in a 1 Hz band surrounding a specified value of frequency.

2.1.18 *Distortion* (Crossover distortion, non linearity, harmonic distortion)

2.1.18.1. Crossover distortion principally relates to the non-linearity or discontinuity generated in class B type output stages when the current sourcing turns off and the current sinking turns on. It is minimised by careful biasing techniques on the chip. It can be observed as a small discontinuity in the output waveform of an amplifier with a sinusoidal input, operating under high closed-loop or open-loop gain configurations. Some more recent amplifier designs are advertised as having negligible crossover distortion. If it is not specified, beware: it should not be inferred that it does not exist. Figure 2.7 shows the effect of crossover distortion on the output waveform.

Crossover distortion is reduced by the application of negative feedback, the higher the loop gain the better the performance.

2.1.18.2. Non-linearity usually becomes more significant as the output swing nears the limit of the output range. It is a rarely defined parameter by op amp manufacturers, but is really very important in the design of transducer amplifiers. Accurate transducers demand amplifiers which do not contribute significant errors to the transducer performance. Transducer non-linearity is defined in many ways, including best straight line, terminal non-linearity and independent non-linearity; careful attention should be given to these definitions, so that the full implications are understood. These limitations are specially important in the situation where interchangeability is

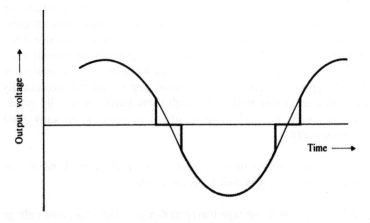

Fig. 2.7 Crossover distortion

required. If linearity is important, there is no substitution for measurement. The objectives in minimising non-linearity would be to restrict output voltage and current swings to well within the operating range, and to maximise loop gain.

2.1.18.3. Harmonic distortion: one manufacturer defines this as:

'Percentage harmonic distortion is 100 × the ratio of the root mean square (rms) sum of the harmonics to the fundamental

$$= \frac{\sqrt{(V_2{}^2 + V_3{}^2 + V_4{}^2 + \ldots)}}{V_1} \times 100 \text{'}$$

2.1.19 Drift with time

Device ageing and resistor drift can contribute to drift with time effects. The rate is usually non-linear with time, often reducing in slope during life. After the initial high drift rate, it slows down and may or may not be accumulative. An approximation to total drift can sometimes be obtained by multiplying the initial drift rate by the square root of the number of days, e.g., if 1 μV/day is specified, total drift after 49 days would be 7 μV.

Although modern high-performance op amps have very low drift with time values, it is likely that chopper or varicap modulated amplifiers will continue to provide the best performance, which results from the shifting of low-frequency information up the band to the carrier frequency.

2.1.20 Channel separation

This is used to describe the crosstalk effects between amplifiers on a multi-amplifier chip, e.g. quad op amps. It is usually defined for a specified frequency range (crosstalk usually deteriorates as frequency increases) as a ratio in dB. A signal is applied to one amplifier and the input referred signal from an adjacent amplifier is measured.

2.1.21 Current source and sink

Recent amplifier improvements include the ability of the output to approach very closely to the extremes of the power supply rails. It is therefore desirable to know the current which the output can supply or sink before saturation of the output devices occurs.

2.1.22 Current mirror gain

Some amplifiers do not have fully differential inputs, and the 'current mirror' technique, where a current into a second terminal is used to control the current in the main input terminal, is utilised. The ratio of the controlled current to the input current is the mirror gain, and is normally near to unity. Also defined is:

2.1.23 Current mirror gain drift

This is the temperature dependence of mirror gain.

2.1.24 Phase margin

This is the extra loop phase shift required to cause the instability in the unity loop again configuratiòn.

2.1.25 Thermal feedback coefficient

This is the resulting input offset voltage error due to increased device dissipation. It is expressed in $\mu V/mW$.

2.2 Protection

The previous section is a brief review of how to approach the initial design requirement, followed by the most frequently encountered op amp parameters and the ways in which they are defined. The important thing to remember is that the same parameter may well be defined differently by different manufacturers, and so the full implications of the values given must be carefully considered. We will now move on to some of the essential considerations regarding device protection.

Many of the early op amps suffered from limitations which contemporary devices do not have. Some of these early devices are still in extensive use, and so it is worth while here to describe these problems. Whenever an unfamiliar device is being utilised, it is always worth while to check that it is protected from these difficulties. It is also necessary to consider the effects of external influences which can damage or destroy any device by gross overstressing.

2.2.1 Latch-up

This is a problem usually encountered when amplifiers are operating under low closed-loop gain, high signal-level applications. The output

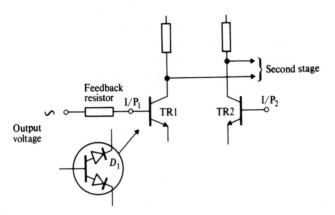

Fig. 2.8 Amplifier input stage saturation

voltage swings to its limit (see Fig. 2.8) and the collector of TR 1 bottoms. However, the base becomes overdriven, so that the collector-base diode D_1, normally reverse biased, becomes forward biased and the transistor no longer operates in a linear inverting mode. This in turn means that the feedback around the loop is positive rather than negative. This then becomes regenerative and so the loop 'latches up' at one limit and stays there. The amplifier can only be reset into its linear operating range by switching off the power supply. For amplifiers where there is a danger of this problem, the method of protection is described in the manufacturers' literature.

2.2.2 Output protection

There are three common forms of circuit damage or destruction at the output terminal.

2.2.2.1. Dissipation limits in the output devices can be exceeded, in which case the effect is dependent on the thermal time constant and thermal resistance. Some early devices have partial internal protection, in which case the permitted output short-circuit duration is for a defined time, perhaps 5 seconds. Later devices are protected from continuous short-circuits, but the maximum permissible internal dissipation rating must not be exceeded. For example, the output current limit may be 25 mA. At a supply voltage of 15 V this implies an internal dissipation of $12 \times 25 = 300$ mW, assuming that the output drive transistor saturates at a collector-emitter voltage of 2.3 V, and 0.7 V is dropped across a current sense resistor. Now a typical package dissipation is 500 mW at 25 °C ambient. This corresponds with:

T_{jmax} (maximum junction temperature) = 100 °C
Q_{j-a} (thermal resistance, junction to ambient) = 150 °C/W
i.e. at 25 °C. P_{tot} (maximum power dissipation)

$$= \frac{T_{jmax} - T_{amb}}{Q_{j-a}} = 500 \text{ mW}$$

The temperature at which a package dissipation of 300 mW is permitted would be

$$T_{jmax} - Q_{j-a}(300 \text{ m W}) = 55° \text{ C}$$

In this case, then, continuous short-circuits would not be permissible above an ambient temperature of 55 °C.

2.2.2.2. Secondary breakdown of output devices is another possible failure mode. This is a form of avalanche breakdown which occurs at specific combinations of current and voltage, and causes irreversible destruction at very high speed. The device may well be within a safe dissipation limit, and yet still suffer from secondary breakdown. This occurrence is unusual in devices with short-circuit protection.

Improved protection is afforded by an external limiting resistor as shown in Fig. 2.9. Resistor R is chosen to limit the output current and internal dissipation to a safe value at all operating temperatures. It is enclosed within the feedback loop so that the output resistance still remains low, the effective R being reduced by the loop gain.

2.2.2.3. External influences can sometimes damage the output devices of op amps. In one case known to the author, the output wires from the amplifier were routed parallel to control cables which occasionally carried large transient currents. The electromagnetic field created produced voltage transients in the output terminals of the op amp of sufficient energy magnitude to literally blow up the

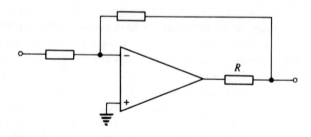

Fig. 2.9 Output protection

output transistors, presumably by reverse biasing. When the chip was extracted from the can, and viewed under a microscope, the damage was clearly visible. This problem was overcome by inserting a voltage- and frequency-limiting filter in the op amp output terminals, as in Fig. 2.10. The effects of C_1 on the loop stability of the amplifier will need to be carefully considered, but the resulting phase lag introduced can be offset by introducing a phase advance by the use of C_2. Also, resistors R_1 and R_2 will limit the maximum output drive voltage available, depending on the magnitude of the load current.

2.2.3 Over voltage protection

In situations where op amps are operated from unstabilised or poorly regulated power supplies, especially in electrically noisy environments, it is often desirable to protect the device from high-speed transient overload on the supply rails. Well-designed supplies for such applications will incorporate SCR (silicon controlled rectifier) 'crowbar' overvoltage devices, so that expensive circuits can be protected from input supply surges and possible series regulator failure. Alternatively, VDRs (voltage dependent resistors) can be used for transient absorption, but it is invariably better to eliminate transients at source whenever possible. Care must be taken, of course, that the zero voltage reference has a low impedance, otherwise the transients may be common to all lines; in this case, any form of filtering or protection referred to the 0 V line will be useless, and a good earth

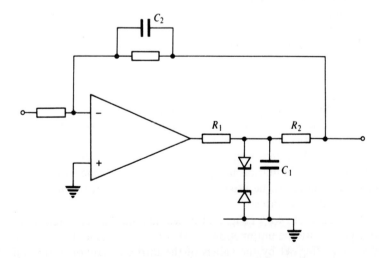

Fig. 2.10 Externally induced transient protection

return is imperative. Field instrumentation for permanent installation may well require protection against line-induced transients from lightning. This is a difficult problem because of the high energies involved, but the usual technique is to use gas discharge tubes directly on the field line, and progressively attenuate the transients from the level of several hundred volts of the tube breakdown using voltage-dependent resistors and zener diodes. Special tubes for this purpose are available on the market, and the techniques used on telephone cables may be applicable. More information on this topic is contributed by Marshall.[4]

2.2.4 Electromagnetic radiation

It is often found that low-level circuits are affected by communications sets in use nearby. This can produce d.c. output shifts even though the normal operating bandwidth of the circuit is well below the transmitting frequency being used. This occurs because the high-frequency RF radiation is detected and rectified by non-linear devices in the circuit, which then produces changes in the bias conditions. The problem can be minimised by:

1. Enclosing the circuit completely within a shielded box.
2. Isolating all wires entering and leaving the box by feed-through capacitors (some of which have integral inductors to act as RF filters), and filtering each line. (The wires leaving and entering the box behave like aerials.)
3. Filling any cracks between mating metal parts of the enclosure with conductive RF gaskets.
4. Using screened wires wherever possible.

2.2.5 Input protection

Most modern integrated-circuit op amps can be operated without damage with either or both input terminals at any level between the two supply rails, but reduced limits are applicable to older types. In cases where the input terminals could go outside the supplies, some sort of voltage limitation using a combination of zener diodes and ordinary diodes can be used; alternatively, the arrangements in Fig. 2.11 may be applicable.

The circuit shown in Fig. 2.11(a) will maintain a very high input resistance until the unijunction transistor(UJT) threshold is reached, the level being set by the values of the intrinsic standoff ratio and resistors R_1 and R_2. Resistor R_3 limits the current after the UJT has

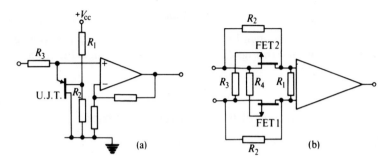

Fig. 2.11 Input protection

fired. This arrangement will protect against positive transients, and negative transient protection can be applied by using a complementary UJT. Circuit 2.11(b) will protect the inputs against large differential input voltages. When the differential input voltage becomes larger than the FET pinch-off voltage, the drain to source resistance becomes very large and the input to the amplifier is limited to the ratio of R_1 to $R_1 + R_2$ times the differential input. Resistors R_3 and R_4 prevent large currents from flowing when the gate source junctions are forward biased.

Protection of junction FET input amplifiers can often be accomplished simply by using large series resistance values in the input terminals. If the input voltage is of sufficient magnitude to break down the reverse gate-channel diode, the current which flows is limited to a very low value by the input resistor and does not cause permanent failure of the input devices. Care will be required in choosing compromise resistor values which do not significantly degrade noise and drift performance.

2.3 Non-ideal parameters

Many of the terms specified in section 2.1 will cause some sort of performance limitation in circuits with demanding requirements. In that sense, perhaps the heading of this paragraph demands a more thorough investigation of the effects of all these problems. However, there are really three fundamental departures from the ideal which are usually selected for special attention, since they tend to occur more frequently in considerations of accuracy. These are open-loop gain, input resistance and output resistance. As will be seen, the effects of the last two are largely controlled by the former, so let us begin with the effects of a finite value of open-loop gain.

2.3.1 Gain

Section 2.1.1 defines open-loop voltage gain. Closed-loop gain can be derived as follows. In Fig. 2.12(a) the circle is a voltage-differencing device, such that

$$V_1 = V_{in} - \beta V_o,$$

where β is the fraction of the output voltage fed back.

Then
$$V_o = AV_1 = A(V_{in} - \beta V_o)$$

and gain
$$G = \frac{V_o}{V_{in}} = \frac{A}{1 + \beta A} \qquad \dots [16]$$

In Fig. 2.12(b), the usual inverting op amp configuration, V_1 is the *effective* input signal to an amplifier with feedback applied (see section 1.8.3). Here $\beta = R_1/(R_1 + R_2)$ and $V_1 = V_{in} R_2/(R_1 + R_2)$, assuming that the input resistance R_{in} is infinite, and the output resistance R_0 is zero. Hence closed-loop gain G, from [16], is given by

$$G = \frac{R_2}{R_1 + R_2} \cdot \frac{A}{1 + A R_1/(R_1 + R_2)} = \frac{R_2}{(R_1 + R_2)/A + R_1} \qquad \dots [17]$$

If A is infinite, $G = R_2/R_1$, the usual expression. It is important here to note a common error. The summing junction x has a signal of V_o/A imposed upon it. This clearly must be so; as A tends to infinity, then V_{error} tends to zero, and that is why the junction is often termed 'virtual earth', in that it is *effectively* at earth potential. The input resistance to V_{in} must then be R_1. In practice, however large the value of A, there is always a very low-level signal present at x, because if there was not, connecting a wire from x to ground would not have an effect! Generally, attempting a measurement at point x will prove to be meaningless and should be avoided. To reiterate, a virtual earth point carries the error signal.

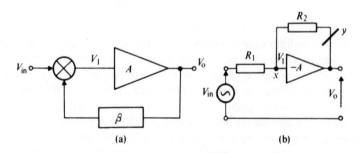

Fig. 2.12 Op amp as a voltage feedback amplifier

From [17], the effects of a finite value of A are obvious. The small gain error resulting from this can be easily eliminated by a small adjustment to the value of R_2, so this is not a great problem. Of far greater significance is the gain stability resulting from variation of A with time, temperature, and power supply. Now from [16],

$$G = \frac{A}{1 + \beta A} \quad \text{where } G = \text{closed-loop gain}$$

$$A = \text{open-loop gain}$$
$$\beta = \text{feedback fraction (constant)}$$

$$\therefore \frac{dG}{dA} = \frac{1}{(1 + \beta A)^2}$$

$$\frac{dG}{G} = \frac{1}{1 + \beta A} \frac{dA}{A}$$

i.e., percentage change of $G = \dfrac{1}{1 + \beta A} \times$ percentage change of A

$$\dots [18]$$

Here it is convenient to define a third gain parameter which often causes confusion. If the feedback loop is broken at Y (Fig. 2.12(b)) and a voltage source is injected, the gain around the loop can be seen to be

$$\frac{R_1}{R_1 + R_2} \cdot A = \beta A$$

This is known as the *loop gain*. Notice that the important factor in [18] which determines stability of the closed-loop gain is βA, the loop gain. The objective is to maximise the loop gain to obtain minimum variations of G with variations of A. From [16], if A is infinite, $G = 1/\beta$.

Absolute closed-loop gain error is then

$$\frac{1}{\beta} - \frac{A}{1 + \beta A} = \frac{1}{\beta(1 + \beta A)} \qquad \dots [19]$$

For the inverter circuit of Fig 2.12(b), $\beta = R_1/(R_1 + R_2)$ and the gain equation is multiplied by the input attenuating constant $K = R_2/(R_1 + R_2)$ (see [17]).
Substituting in [19],

Closed-loop gain error

$$E = \frac{R_2}{R_1\left(1 + \dfrac{AR_1}{R_1 + R_2}\right)} \qquad \ldots [20]$$

(This can be checked by assuming that ideal gain is R_2/R_1 and calculating $R_2/R_1 - [17]$.)

From [18], the stability factor for the inverter is

$$S = \frac{K}{1 + \beta A} = \frac{R_2}{R_2 + R_1(1 + A)} \qquad \ldots [21]$$

Similar expressions can be derived for the non-inverting amplifier Fig. 2.15, assuming $R_{\text{in}} = \infty$ and $R_0 = 0$

The ideal gain here is

$$\frac{1}{\beta} = \frac{R_1 + R_2}{R_1} \qquad \ldots [22]$$

and the actual closed-loop gain from [16] is

$$G = \frac{A(R_1 + R_2)}{R_2 + R_1(1 + A)} \qquad \ldots [23]$$

The closed-loop gain error from [19] is

$$E = \frac{R_1 + R_2}{R_1\left(1 + \dfrac{R_1}{R_1 + R_2}A\right)} \qquad \ldots [24]$$

and the stability factor

$$S = \frac{R_1 + R_2}{R_2 + R_1(1 + A)} \qquad \ldots [25]$$

Open-loop gain of op amps usually tends to decrease with temperature, and to increase with increasing supply voltages. However, recent advances using MOS and bipolar devices on the same chip have been capable of producing amplifiers with negligible change of open-loop gain with voltage. There are often changes of slope of the temperature-dependence curve, and some notable exceptions to the voltage dependence; for example, the open-loop gain of the LM118 actually increases with decreasing supply voltage.

2.3.2 Input and output resistance effects

We will now proceed to examine the effects of the input and output resistance on the performance of an amplifier. Note that the

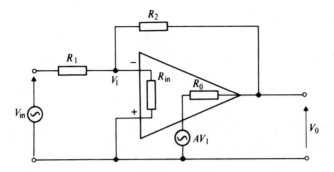

Fig. 2.13 Inverting input and output resistance effects on gain

magnitude of the open-loop gain of some amplifiers is determined by the magnitude of the external loading resistors. This must also include the feedback components because they load the output. Figure 2.13 shows the equivalent circuit. Proceeding with the analysis as in the previous section, the equivalent circuit for calculating β is as shown in Fig. 2.14(a). In this case,

$$\beta = \frac{R_a}{R_2 + R_a}, \qquad \text{where } R_a = \frac{R_1 R_{in}}{R_1 + R_{in}}$$

Open-loop gain is effectively reduced from A to

$$\frac{R_a + R_2}{R_a + R_2 + R_0} \times A,$$

as can be seen from Fig. 2.14(b). The effective input signal to the amplifier from Fig. 2.14(c) is

$$\frac{R_b}{R_1 + R_b} V_{in}, \qquad \text{where } R_b = \frac{(R_2 + R_0)R_{in}}{R_2 + R_0 + R_{in}}$$

(assuming negligible source resistance)

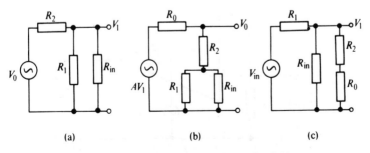

(a)   (b)   (c)

Fig. 2.14 Equivalent networks for Fig. 2.13

Closed-loop gain

$$G = \frac{R_b}{R_1 + R_b} \times \frac{A(R_a + R_2)/(R_a + R_2 + R_0)}{1 + [R_a/(R_2 + R_a)][(R_a + R_2)/(R_a + R_2 + R_0)]A}$$

$$\therefore\ G = \frac{R_b}{R_1 + R_b} \times \frac{A(R_a + R_2)}{R_2 + R_0 + R_a(1 + A)} \qquad \ldots [26]$$

It will be found that the value of R_0 is rarely quoted for amplifiers, and is usually quite small compared with the loading resistors. Open-loop gain is often specified with a defined load, so this should be considered when calculating the effective value. The output resistance will only apply at low frequencies, since it tends to be partially complex and usually increases with frequency.

A typical value for R_{in} (bipolar inputs) might be 10 MΩ, but FET input amplifiers having values in excess of 1 TΩ are now available, and in these cases the effects of R_{in} on the closed-loop gain can usually be ignored. Typical values for R_0 are in the 50–150 Ω region.

Figure 2.15 shows the connection for a non-inverting amplifier. The input signal is a voltage source; if significant source resistance is present, the effect of R_{in} on the input signal will need to be considered (with the inverting configuration, source resistance is easily added directly to R_1). In the above case the effective input signal is V_{in}, and the effective open-loop gain is

$$A_{OL} = \frac{R_2 + R_a}{R_2 + R_a + R_0}A, \qquad R_a = \frac{R_1 R_{in}}{R_1 + R_{in}}$$

$$\beta = \frac{R_a}{R_a + R_2}$$

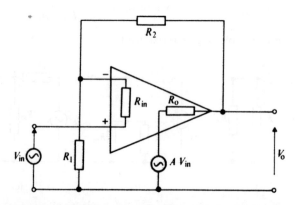

Fig. 2.15 Non-inverting input and output resistance effects on gain

∴ closed-loop gain

$$G = \frac{A(R_a + R_2)}{R_2 + R_0 + R_a(1 + A)} \qquad \ldots [27]$$

If A is very large, $\quad G = \dfrac{R_a + R_2}{R_a}, \quad$ and if $R_{in} \gg R_1$,

$$G = \frac{R_1 + R_2}{R_1} = \frac{1}{\beta}, \quad \text{the idealised value.}$$

In summary, Table 2.1 shows how the values compare. They are computed using the following:

$$A = 50\,000$$
$$R_{in} = 10\,M\Omega$$
$$R_0 = 100\,\Omega$$
$$R_1 = 1k\Omega$$
$$R_2 = 1M\Omega$$

The conclusions to be drawn from Table 2.1 are as follows:

1. Finite open-loop gain has a significant effect on closed-loop gain. In practice this is not a problem since resistors are never perfect and it will always be necessary to include adjustments for accurate gain setting. However, carrying out the calculation enables one to decide on the range of adjustment that is required. Chosen resistor tolerances must be considered of course, e.g.,

$$\text{if } G = \frac{R_2}{R_1}, \text{ then } \frac{\delta G}{G} = \frac{\partial G}{\partial R_1}\frac{\delta R_1}{G} + \frac{\partial G}{\partial R_2}\frac{\delta R_2}{G}$$

and percentage change of $G =$ percentage change of R_2 − percentage change of R_1, so if R_2 has a tolerance of $+2$ per cent and R_1 a tolerance of -2 per cent, then G could be in error by $+4$ per cent from resistor tolerances alone.
2. The closed-loop gains of the inverting and non-inverting configurations are slightly different using the same component values.
3. The effects of typical values of R_{in} and R_0 are negligible at low frequencies.
4. Gain stability factor is a very important parameter. For example, if the open-loop gain of the example amplifier changed by -6 dB (e.g. from 94 dB to 88 dB, a change of 50 per cent), then the resulting change of closed loop gain in this case would be $0.019\,61 \times 50 = 0.98$ per cent.
5. Remember that the effective input and output resistances are modified by feedback as described in the next section.

Table 2.1 *Direct current gain of operational amplifiers showing the effects of finite amplifier parameters.*

Parameter	Inverting amplifier	Value	Non-inverting amplifier	Value
Ideal gain	R_2/R_1	1 000	$\dfrac{R_1+R_2}{R_1} = \dfrac{1}{\beta}$	1 001
Gain with finite A	$R_2\Big/\left(\dfrac{R_1+R_2}{A}+R_1\right)$	980.373	$\dfrac{R_1+R_2}{\dfrac{R_1+R_2}{A}+R_1} = \dfrac{A}{1+\beta A}$	981.353
Gain error factor	$R_2\Big/R_1\left(1+\dfrac{AR_1}{R_1+R_2}\right)$	19.627	$\dfrac{R_1+R_2}{R_1\left(1+\dfrac{AR_1}{R_1+R_2}\right)}$	19.647
Gain stability factor	$R_2/(R_2+R_1(1+A))$	0.01961	$\dfrac{R_1+R_2}{R_2+R_1(1+A)}$	0.01963
Gain when the effects of R_{in} and R_0 are considered.	$\dfrac{R_b}{R_1+R_b}\cdot\dfrac{A(R_a+R_2)}{R_2+R_0+R_a(1+A)}$	980.369	$\dfrac{A(R_a+R_2)}{R_2+R_0+R_a(1+A)}$	981.448

2.3.3 The effects of feedback on terminal resistances

The application of feedback to an amplifier modifies the input and output terminal impedances presented to the outside world. This is a very important effect which is often utilised in practical circuits. We have already discussed the 'virtual earth' concept (section 2.3.1), and have shown that the input resistance to the inverting amplifier is approximately the value of the input resistor R_1. The input resistance to the non-inverting amplifier can be obtained with reference to Fig. 2.16.

The equivalent input circuit from (a) is shown at (b) and (c) (assuming that $\dfrac{R_1 R_2}{R_1 + R_2} \ll R_{in}$). $V_o = \dfrac{A}{1 + \beta A} V_{in}$.

The signal appearing at the inverting input is

$$\beta V_o = \beta \frac{A}{1 + \beta A} V_{in}$$

From (b) and (c),

$$i_{in} = \frac{V_{in}\left(1 - \dfrac{\beta A}{1 + \beta A}\right)}{R_{in}} = \frac{V_{in}}{R_e}$$

Where R_e is the effective input resistance

$$\therefore R_e = \frac{R_{in}}{1 - \dfrac{\beta A}{1 + \beta A}} = (1 + \beta A)R_{in} \qquad \dots [28]$$

Again, the loop gain is the important factor, and the effective input resistance is much higher than the fundamental amplifier resistance specification R_{in}. The frequently used application of this property is

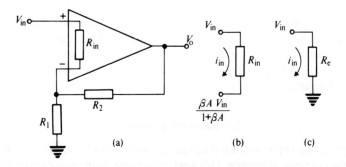

(a)   (b)   (c)

Fig. 2.16 The effect of feedback on input resistance

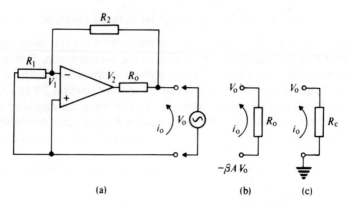

Fig. 2.17 The effect of feedback on output resistance

the voltage follower where total feedback is applied and $\beta = 1$ (i.e., $R_1 = \infty$, $R_2 = 0$ above). Very high effective input resistances can be obtained with modest R_{in} specifications.

Figure 2.17(a) shows the method for evaluating the effective output resistance R_C of an amplifier with feedback. V_o is a voltage generator connected to the amplifier output, and the input is returned to the common as would be the case with a voltage source. Then $V_1 = \beta V_o$ and $V_2 = -\beta A V_o$. The equivalent circuit of the output is given in Fig. 2.17(b) and (c).

$$i_0 = \frac{V_o(1 + \beta A)}{R_0} = \frac{V_o}{R_C}$$

$$\therefore R_C = \frac{R_0}{1 + \beta A} \qquad \dots [29]$$

The effective output resistance is very much lower than the open-loop value R_0. The same value is obtained from the non-inverting configuration, and if βA is large then the effective output resistance can be very low. As the frequency increases, R_C will increase due to the reduction of open-loop gain with frequency, and a complete examination requires the consideration of the imaginary part of A (see section 3.9). The effective input resistance will reduce as A decreases with frequency.

2.4 Earthing problems

In precision applications it is of vital importance to ensure the quality of the common reference. This applies to noise from sources external to the circuit and to earth loop currents. A common problem is noise

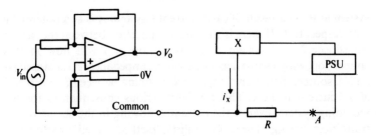

Fig. 2.18 The effects of poor earthing

induced by currents flowing in other parts of the same circuit. Figure 2.18 shows how the difficulty arises. R is the resistance of the common line return to the power supply, and X is another part of the circuit excited from the same supply. Current i_x flows through R and produces a volt drop which is superimposed on the common of the amplifier. This can be equivalent to producing a common-mode input error to the amplifier resulting in error generation. Since common-mode rejection ratio is frequency-dependent, and the current may well be transient in nature, large differential errors can result. The solution is to take the common lines of any parts of the circuit likely to be carrying significant currents, especially from switching circuits, separately direct to A, so that they do not flow through the common resistance R. This is often easy to implement at the design stage, but far more difficult to correct after manufacture.

Another very commom source of trouble is earth loop currents, especially in transducer applications. These are caused by the connection of different parts of a system at differing potentials with respect to ground. The existence of differing potentials is often caused by the same process as is described above, but this time with respect to the mains earth reference. Consider Fig. 2.19 for example; transducer T is monitoring a process variable and is connected to a recording

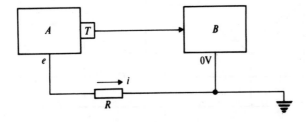

Fig. 2.19 Earth loop currents

system at **B**. As a result of earth current i, system A is at a potential e with respect to B. If there is any direct connection between the case of transducer T and system B, a proportion of i will flow in this connection, with a value dependent on the impedance. If i is caused by an a.c. source, then even a capacitive coupling between T and B will allow current to flow. This can have serious consequences for the recording system, even if the current is flowing in the screen of the transducer wiring. The only realistic method of eliminating this problem is to introduce an isolator between T and B. This may be by transformer coupling or optical isolation. It also means that the power supply for T must be isolated from the power supply to the recording system.

2.5 Power supply tracking

We have mentioned in the definition of terms in section 2.1.10 the effects of power supply rejection ratio, and the advantages of using tracking rails. This is especially valuable in bridge transducer applications such as indicated in Fig. 2.20. For example, if $+ V_{cc} = - V_{cc}$, then under ideal conditions the bridge output terminals are at 0 V. If $+ V_{cc}$ increases by the same amount as $- V_{cc}$ decreases, then the common-mode voltage remains at 0 V, optimising the common-mode rejection of the amplifier. If the amplifier is also connected to the same supplies, the input terminals remain at the same levels with respect to the internal devices. It is possible to obtain dual tracking voltage regulators as single devices, but the basic arrangement is usually as

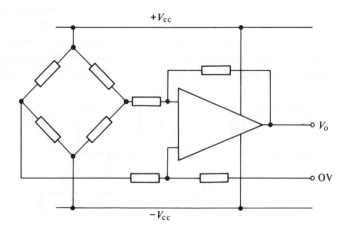

Fig. 2.20 Tracking power supplies and a bridge amplifier

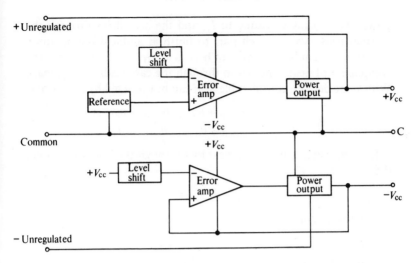

Fig. 2.21 Block diagram of a typical tracking voltage regulator

shown in Fig. 2.21 for those interested in constructing their own. A single stable reference source is used for the derivation of both rails, so that any variations with temperature or time are superimposed equally on both.

In practical situations the common-mode output of the transducer may change with temperature due to thermal mismatch in the bridge. One possibility for overcoming this problem is to detect the common-mode voltage and use it to modify the supply rails in such a way that the original levels are restored.

2.6 Thermal response

We have briefly mentioned thermal resistance in section 2.2.2.1. The equivalent circuit can be drawn as in Fig. 2.22. The device dissipates

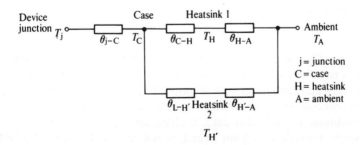

Fig. 2.22 Electrical equivalent of thermal effects

power which causes heating to T_j, and the heat flows through the thermal resistances of each part to finish at ambient temperature. The extra unfamiliar component here is $\theta_{L-H'}$ which represents the component of heat conduction through the case leads or terminals. With some devices, considerable heat can be extracted by this means.

$$\theta_{j-A} = \theta_{j-C} + \frac{(\theta_{C-H} + \theta_{H-A})(\theta_{L-H'} + \theta_{H'-A})}{\theta_{C-H} + \theta_{H-A} + \theta_{L-H'} + \theta_{H'-A}} \quad \ldots [30]$$

If no heatsink is used, and heat loss through the leads is ignored, then

$$\theta_{j-A} = \theta_{j-C} + \theta_{C-A}$$

The maximum dissipation then, is given by

$$P_{\text{tot max}} = \frac{T_{j\text{max}} - T_{\text{amb}}}{\theta_{j-A}} \quad \ldots [31]$$

Maximum device temperature may vary from 85 to 150 °C.

Now this dissipation figure represents the average steady-state dissipation, but in some applications the transient dissipation becomes relevant. Power op amps can often drive high pulsed currents where the peak power is much greater than the average power. It becomes relevent in such circumstances to ask what effect the thermal time constant has on the effective averaging of the power dissipation. Where multiple amplifier chips are packaged in a single unit, the maximum permitted power dissipation at a device temperature is usually the total for all amplifiers; the dissipation for each will then depend on how many amplifiers are being used. Sometimes a device may be operating in a situation where the thermal resistance between case and ambient is unknown, and it is difficult to be sure that the maximum permitted junction temperature is not being exceeded. In this instance the operating case temperature can be measured by a surface-sensing thermocouple probe at the maximum expected ambient temperature. Then

$$T_j = T_C + P_{\text{tot}}\theta_{j-C} \quad \ldots [32]$$

Secondary breakdown can cause device failure in situations where it is not dissipation-limited. This occurs at certain combinations of voltage and current, usually in output transistors, and is destructive. As collector-to-emitter voltage is increased for any specific base-drive condition, the collector current characteristic first enters the avalanche breakdown region (which is not destructive under limited current conditions) and then further increases of current occur until secondary breakdown is reached. The collector–emitter voltage then falls very rapidly to a low value, usually destroying the device. The usual explanation of this effect is to assume that the current distribution within the device itself becomes uneven, which then produces local overheated areas. This has the effect of tending to increase still further the conduction at that point due to the fact that the V_{BE} of a transistor decreases in magnitude with temperature. The result is a positive feedback, resulting in higher and higher current densities in one part of the semiconductor material. More complete discussions of this problem are to be found in manufacturers' literature.

Finally in this chapter, a word of warning. Pay attention to detail with any linear circuit design. Things are not always what they seem to be! The author has had experience of this when it was discovered that a small number of amplifiers in a production system were oscillating, even though the recommended frequency compensation arrangement had been utilised, and even though many other amplifiers were stable when operating in the same circuits. It was eventually discovered that the device case had actually been incorrectly marked, so that the chip inside the case did not correspond with the external marking!

3

Frequency-dependent parameters

3.1 Frequency compensation

This is an extensive subject, and is rigorously discussed elsewhere.[5] I therefore intend to limit consideration here to some of the practical aspects of the problem, and perhaps mention some topics which are not often discussed in the literature.

3.1.1 Why frequency compensation?

In fact, this is something of a misnomer, since the usual interpretation is 'the shaping of an amplifier's gain with frequency plot, thereby procuring a definable phase shift and loop gain at frequency f'. Basically, for all op amps, if the open-loop gain is measured and plotted against a frequency axis, there will be a gain reduction with increasing frequency. This can be explained by the inevitable presence of various stray and feedback capacitances associated with both active and passive components on the chip. In a practical application there are also stray capacitances associated with the device input pins and the wiring connected to them. It is therefore imperative to take great care with the physical layout of high-frequency amplifiers if predictable characteristics are to be obtained. A very small capacitance, for example, 10 pF, has an impedance of 15.9 KΩ at 1 MHz, which could have a very large effect on external resistance values.

Since the reduction of open-loop gain with frequency is dependent on reactive impedances, there is bound to be an associated phase shifting effect. For a simple single $R - C$ circuit, the rate of amplitude attenuation with frequency has an ultimate roll-off rate of 6 dB/8^{ve} or 20 dB/10^{ade} (i.e. the amplitude halves for every doubling of the frequency). Associated with this, the transfer function shows us that the maximum phase shift is 90°. Chart 3.1 shows the relative amplitude and phase response for several buffered R–C stages, and

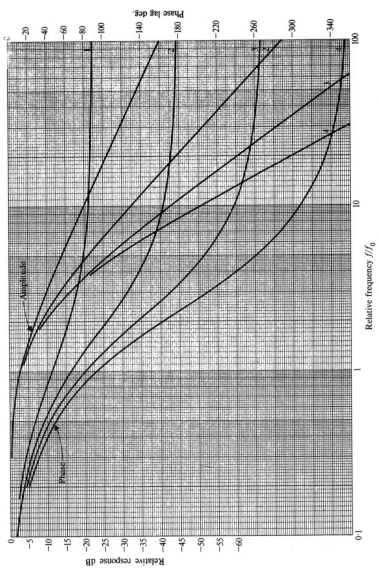

Chart 3.1 Relative response of buffered R–C sections

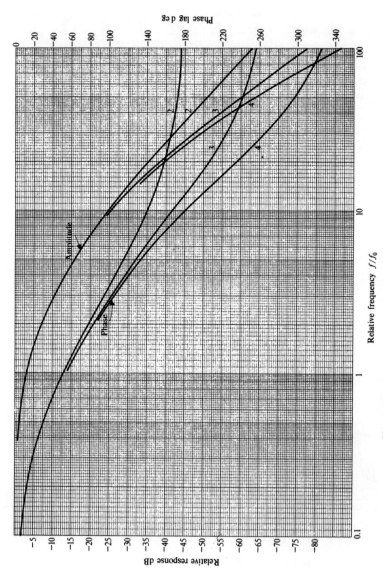

Chart 3.2 Relative response of cascaded R–C sections

Chart 3.2 the same parameters for an equal-valued ladder where each successive stage loads the previous one. The response curves for the open-loop gain of an op amp tends to be a complex addition of these characteristics, but the actual responses can be drawn in plots called 'Bode' plots as in Fig. 3.1. The curves are named after Hendrik W. Bode, who published a paper in the *Bell System Technical Journal* and subsequently a book on feedback amplifier design in 1945.[6] In practice, the − 3 dB points of the various parts of the circuit producing the roll-off do not all occur at the same frequency.

It might therefore seem to be a reasonable proposition to infer a certain value of phase shift resulting from a known rate of attenuation in the magnitude response. This is valid for simple networks, but is not always applicable to complex situations such as exist in op amp characteristics. This basic problem has been further discussed by Thomas.[7] However, for most frequency-compensation requirements, it is usually adequate to associate a given rate of roll-off in the amplitude response with a specific open-loop phase shift.

Chart 3.1 shows the curves for the situation where the time constant of each successive circuit is identical. The usual situation with op amps (uncompensated) is that the roll-off is determined by the equivalent of several buffered R − C sections isolated from each other, and each with differing time constants separated by significant amounts. Hence

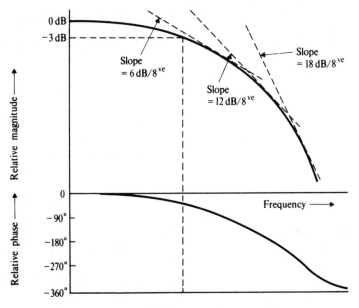

Fig. 3.1 Bode plots for amplifier frequency response

to produce a theoretically equivalent curve using the calculations shown in Appendix 1(F55), it is necessary to use different time constants for each section.

The importance of phase shift becomes very clear when we consider what happens in an amplifier with feedback applied. The feedback is normally connected in a negative sense, that is, it subtracts from the input producing a small error voltage. As frequency increases, the relative phase shift around the loop changes, and eventually at 180 ° the feedback signal is now in phase with the input. Consequently it adds to the input signal, producing a positive feedback situation and hence instability or oscillation. It is therefore essential to have advance knowledge of the open-loop amplifier characteristic before closing the loop, to ensure an adequate margin of stability.

3.1.2 Compensation methods

There are various methods employed for producing the desired frequency response characteristic, including networks which modify the magnitude and the phase response; it is often difficult to modify one without affecting the other, although some possible methods exist as shown in Appendix 3(f 56). There are three basic techniques used to modify the open-loop gain response, thereby procuring phase-shift values unlikely to cause instability.

Figure 3.2 shows the simplest method, which is almost invariably used for internally compensated amplifiers, as only a single time constant is required. The amplifier is forced to roll off on a single R–C time constant from a − 3 dB point of a very low frequency, usually about 10 Hz. This ensures that the open-loop gain has fallen to unity with the phase shift dominated by a single time constant and thus

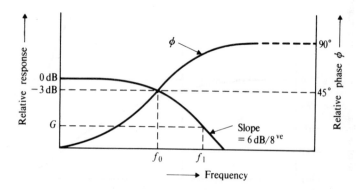

Fig. 3.2 Single time constant frequency compensation

having a maximum of somewhat greater than 90 ° (there is bound to be some phase contribution from the rest of the amplifier). It is clear, therefore, that if the loop is closed to produce gain G, the closed-loop response cuts the open-loop response at a point where the rate of roll off is -6 db/8^{ve}, implying 90 ° phase shift and unconditional stability. This is the simple test usually adopted – does the closed-loop line cut the open-loop line either where the slope is considerably less than 12 dB/8^{ve} (i.e. 180 ° phase shift and the danger point) or where the loop gain is less than unity? If it does, then we can be reasonably confident that the amplifier will be stable. Assessing the margin of stability is another matter, however, and is discussed in Chapter 6. The fact that the open-loop gain, -3 dB frequency is as low as 10 Hz is often not important in low-frequency applications. The magnitude of the open-loop gain is so high at d.c. that there is still adequate bandwidth at a closed-loop gain of G. For example, assume that -3 dB is at 10 Hz, that roll-off beyond 10 Hz is a constant 6 dB/8^{ve}, and that d.c. open-loop gain at f_0 is 100 dB. The frequency f_1 at a closed-loop gain of $G = 100$ is then 10 kHz.

Figure 3.3 shows the method of compensation used for externally compensated amplifiers and where improved bandwidth is required. Breakpoints are introduced into the open-loop response and then successively removed as the amplifier's own open-loop roll-off begins to take effect and produce increasing phase shift. Finally, phase advance is introduced to offset the effects of the internally increasing phase lag. The calculations will not be discussed in detail here since they are readily found elsewhere in the literature and in

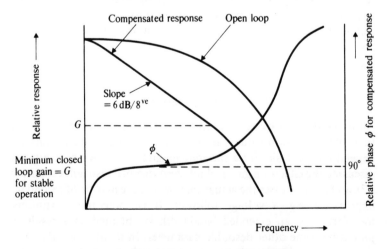

Fig. 3.3 Externally compensated amplifier

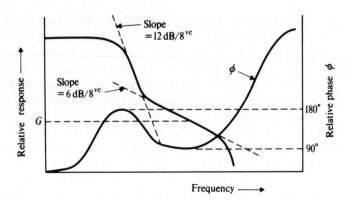

Fig. 3.4 Compensation for bandwidth extension

manufacturers' application information. The detailed calculation of a complex frequency-compensation technique invariably requires a knowledge of the source impedances of the compensation terminals, and if standard recommended networks are not suitable, then recourse must be made to the manufacturer to obtain typical internal circuit values. Occasionally the values can be obtained from a circuit diagram of the integrated circuit, but great care must be exercised in the use of these, as the values shown are nominal and can be subject to quite large tolerances.

Occasionally, wider bandwidth at low closed-loop gains can be obtained by deliberately allowing (or causing) the open-loop response to roll off at a high rate initially, and then by inserting phase advance to reduce the rate and hence the phase shift to acceptable levels when the loop is closed at G. Figure 3.4 shows the effect of this compensation. It is clear that if the loop is closed at higher gain levels then instability can occur.

3.2 Phase linearity

Although in many cases the loop is closed at a gain very much lower than the open-loop value, the reasons for doing so do not always depend on the desire for a specified magnification. We have already discussed the effects on gain stability and terminal resistances (section 2.3), and in these cases the actual gain magnitude may be of secondary importance, e.g. a voltage follower. It is sometimes fortuitous, therefore, that an extended bandwidth is obtained as a result of procuring some other desirable parameter. In many cases, the wide bandwidth can be an embarrassment rather than an advantage, since the wider the bandwidth the greater the contribution of noise at higher frequencies to the overall signal-to-noise ratio. In general, the rule is to design only to the bandwidth required, and not exceed it; this must be considered in relation to transient response and phase linearity which can dictate the amplitude response characteristic required. A brief description of what is meant by phase linearity, since many aspiring designers do not fully appreciate the implications, is relevant at this point.

Consider a complex waveform as shown in Fig. 3.5(a), comprised of the addition of waveforms in (b) and (c). For this waveform to be correctly reproduced (i.e. for the two components to be added to each other with the correct phase relationship), each frequency component must be phase-shifted in the system such that the absolute time delay for both is equal (any amplifier with phase lag will insert a time delay). This implies a different value of phase shift for (b) and (c), and leads to the conclusion that phase shift must be directly proportional to frequency.

Let us imagine that point *d* in Fig. 3.5 was delayed in time to point *e*, in such a way that both waveforms still had the same values at *e* as they

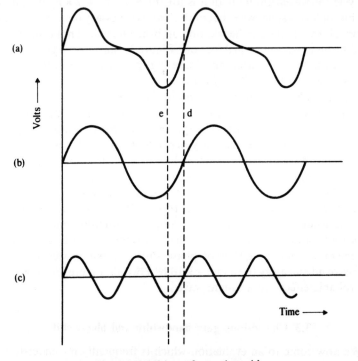

Fig. 3.5 Addition of two sinusoids

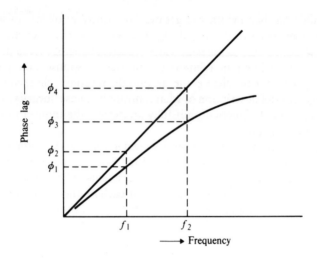

Fig. 3.6 Phase linearity with frequency

did at d. It is clear that the time $d - e$ is a much larger proportion of one cycle for waveform (c) than it is for (b). Since $f_{(c)} = 2 \times f_{(b)}$ the phase shift of (c) must be twice that of (b). This is necessary to reproduce the waveform at (a) since the addition of (b) and (c) would not result in (a) if the times were not equal for both. Figure 3.6 shows graphically what is meant by this reasoning. Phase lag must increase linearly with frequency, so that $f_2/f_1 = \phi_4/\phi_2$.

If the phase characteristic is non-linear, then $f_2/f_1 = \phi_3/\phi_1$ and the resultant waveform is a distorted version of the input. It therefore follows that any network or amplifier which has a non-linear phase response as well as a frequency-dependent amplitude response, will distort complex waveforms as well as affect their amplitude. This is nowhere more obvious than in the response to a square wave, where, apart from rise-time degradation or sag introduction, overshoot and ringing can be caused by phase distortion. If complex waveforms are to be accurately reproduced by an amplifier, attention must be given to the phase response as well as the frequency response; it is well to remember that the effects of phase shift can extend well into the passband and away from the -3 dB bandwidth. Figure 3.7 shows the probable effects on a square wave.

3.3 Closed-loop gain bandwidth and phase shift

We now come to an evaluation which is frequently misunderstood and often leads to erroneous conclusions. Figure 3.8 shows the gain

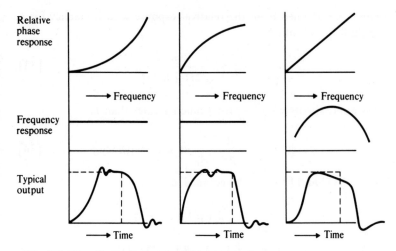

Fig. 3.7 The effects of phase and frequency response on a square wave

magnitude variation with frequency for an amplifier using a single breakpoint frequency-compensation arrangement (the usual technique for internally compensated amplifiers). The -3 dB point is fixed at f_0 and the open-loop gain falls from its low-frequency value of A_0 at a rate approaching 6 dB/8^{ve} beyond f_0. The loop is then closed to produce low-frequency gain G, and the closed-loop bandwidth is defined as f_1. Consider initially the magnitude only.

$$G = \frac{A}{1 + \beta A} \quad \text{and} \quad A = \frac{A_0}{\sqrt{1 + (f/f_0)^2}}$$

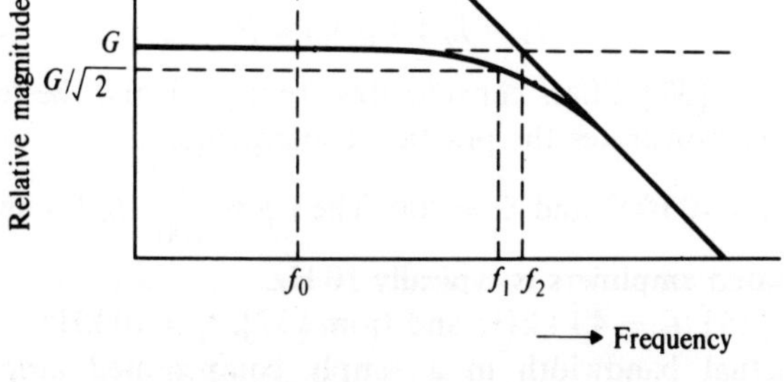

Fig. 3.8 Single breakpoint frequency compensation

where A is defined from the relative response with frequency for a single R–C roll-off. At f_1,

$$|G|_{f_1} = \frac{A_0}{\sqrt{2}(1 + \beta A_0)} \qquad \dots [33]$$

A is approximately $\dfrac{A_0}{(f/f_0)}$, for f much greater than f_0

$$\therefore |G| = \frac{A_0}{(f/f_0) + \beta A_0} \text{ at any frequency } f \qquad \dots [34]$$

At f_1, [33] and [34] can be equated, and

$$\frac{f_1}{f_0} = \sqrt{2} + \beta A_0(\sqrt{2} - 1)$$

$$\therefore f_1 = f_0[\sqrt{2} + \beta A_0(\sqrt{2} - 1)]$$

For large values of loop gain, $f_1 = 0.414 \beta A_0 f_0$ $\qquad \dots [35]$

The foregoing has been calculated using magnitudes; this will be seen below to be erroneous. The correct expression for the variation of A with frequency must include the effects of phase shift, hence

$$A = \frac{A_0}{1 + \mathrm{j}(f/f_0)}$$

$$\therefore G = \frac{A_0}{1 + \mathrm{j}(f/f_0) + \beta A_0} = \frac{A_0}{(1 + \beta A_0) + \mathrm{j}(f/f_0)}$$

Writing magnitudes,

$$|G| = \frac{A_0}{\sqrt{(1 + \beta A_0)^2 + (f/f_0)^2}} \qquad \dots [36]$$

Again equating [36] with [33],

$$2(1 + \beta A_0)^2 = (1 + \beta A_0)^2 + (f_1/f_0)^2 \text{ and}$$

$$f_1 = f_0(1 + \beta A_0) \approx \beta A_0 f_0 \qquad \dots [37]$$

Now clearly [37] differs considerably from [35], and the following example demonstrates the practical consequences.

Let $A_0 = 100\,000$ and $G = 100$. Then $\beta \approx \dfrac{1}{100} \cdot f_0$ for internally compensated amplifiers is typically 10 Hz.

From [35], $f_1 = 4.14 \text{ kHz}$ and from [37], $f_1 = 10 \text{ kHz}$.

The actual bandwidth in a simply compensated amplifier is therefore much nearer to f_2, the crossing of the horizontal projection of G with the open-loop curve, than might otherwise be thought from a simple magnitude examination. Again the relevant factor is the *loop gain*, βA_0. For amplifiers with more complex frequency-compensation networks, the same method can be used, but the second- or third-order expressions would be substituted for the variations of A with frequency. Barker has produced a useful analysis for the optimisation of closed-loop response in amplifiers with two separate breakpoints in the open-loop response.[8] The consequence of this bandwidth extension is that much-improved gain accuracy can be obtained within the passband than would otherwise have been considered possible if the magnitude of A had been directly substituted in the basic gain equation. Continuing the above example, and assuming $\beta = 1/100$,

d.c. gain $G_1 = \dfrac{A_0}{1 + \beta A_0} = 99.9$

Using the magnitude change of A with frequency gives

$$G_2 = \frac{A_0}{\sqrt{1 + (f/f_0)^2} + \beta A_0} = 99.005$$

But using [36], which includes the effects of phase shift,

$$G_3 = \frac{A_0}{\sqrt{(1 + 10^3)^2 + 100}} = 99.895$$

Assuming an ideal gain of 100 (if A_0 was infinite), gain error in each case above is:

$$G_1 = 0.1 \text{ per cent}$$
$$G_2 = 0.995 \text{ per cent}$$
$$G_3 = 0.105 \text{ per cent}$$

The true gain error G_3 at $f = 10f_0$ is therefore only slightly greater than the d.c. error in spite of the fact that A has fallen in magnitude from 100 000 to 9950.

We have seen the effects of the open-loop response on the gain magnitude, and we must now consider the closed-loop phase response.

Open-loop gain $= A = \dfrac{A_0}{1 + j(f/f_0)}, \quad \phi = -\tan^{-1}(f/f_0)$

and as (f/f_0) increases, ϕ tends to $-90°$.

At $f = 10f_0$, $\phi_A = 84.3°$.

Closed-loop gain $G = \dfrac{A_0}{(1 + \beta A_0) + \mathrm{j}\,(f/f_0)}$,

$$\phi = -\tan^{-1} \frac{(f/f_0)}{1 + \beta A_0} \qquad \ldots [38]$$

In the above example, $\phi_G = 0.57°$. The closed-loop phase shift is thus very much reduced from the open-loop value. Chart 3.3 shows how the closed-loop phase shift of a simply compensated amplifier varies with frequency and loop gain.

The above expressions have been derived using the basic closed-loop gain equation for a negative-feedback amplifier, [16]. It was seen in Chapter 2 that this expression can be applied directly to the non-inverting amplifer connection, but is modified in the inverting configuration due to the effective attenuation of the input signal. This means that the bandwidth and phase shift for the two configurations are slightly different, in the same way that the actual gain and gain errors are different as shown in Table 2.1. Generally these differences are small, and the above calculations give an adequate guide to closed-loop performance. If a more rigorous analysis is required, the complex expression for A can be substituted in the expressions for the inverting configuration used in Chapter 2. The bandwidth differences become most apparent at unity closed-loop gain, because for the non-inverter $\beta = 1$ but for the inverting amplifier $\beta = \frac{1}{2}$, so that given identical open-loop responses the non-inverting unity-gain amplifier will have approximately twice the bandwidth of the unity-gain inverter. A further comment on this is to be found in section 6.5.

3.4 Small-signal transient response

It is evident that the closed-loop transient response is largely determined by the open-loop gain and phase characteristics, and an amplifier with two open-loop breakpoints can be compared with the response of a typical second-order system. Such a system can produce peaking in the gain magnitude characteristic and ringing in response to transient inputs. It is possible to introduce a pulse input to such an amplifier, and by measurement of the overshoot and ringing at the output to compute the open- and closed-loop gain responses. A useful derivation of this is given by Huehne.[9] The open-loop response of the simple single-breakpoint amplifier can be written

$$A = \frac{A_0}{1 + \mathrm{j}\omega T} = \frac{A_0}{1 + sT}$$

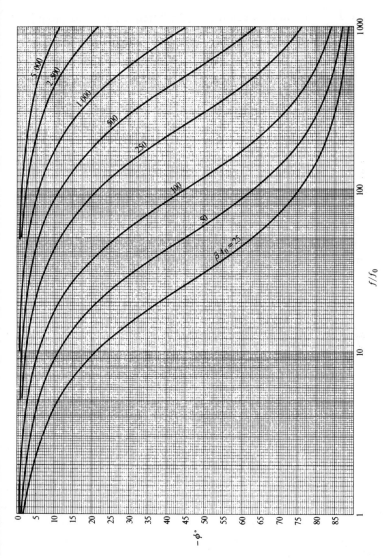

Chart 3.3 Closed loop phase shift – single break point

and the closed-loop gain,

$$G = \frac{A}{1 + \beta A} = \frac{A_0}{(1 + \beta A_0) + sT}$$

where $T = 1/\omega_0$ at the open-loop -3 dB frequency. If the input is a step function $EH(t)$, then the output is

$$\mathscr{L} V_o = \frac{E \cdot A_0}{s[(1 + \beta A_0) + sT]} = \frac{E \cdot A_0/T}{s\left[\dfrac{1 + \beta A_0}{T} + s\right]}$$

$$\therefore V_o = E \cdot \frac{A_0}{T} \cdot \frac{T}{1 + \beta A_0} \left(1 - \exp\left[-(1 + \beta A_0)t/T\right]\right)$$

$$= E \cdot \frac{A_0}{1 + \beta A_0} \left(1 - \exp\left[-(1 + \beta A_0)t/T\right]\right) \quad \ldots [39]$$

The output rise time is $\dfrac{2.2T}{1 + \beta A_0}$

For an amplifier with two breakpoints in its open-loop response, where $T_1 = 1/\omega_1$ and $T_2 = 1/\omega_2$,

$$A = \frac{A_0}{(1 + sT_1)(1 + sT_2)} \quad \text{and} \quad G = \frac{A_0}{(1 + \beta A_0) + s^2 T_1 T_2 + s(T_1 + T_2)}$$

The response to a step input function of $EH(t)$ is

$$\mathscr{L} V_o = \frac{E \cdot A_0}{s[(1 + \beta A_0) + s^2 T_1 T_2 + s(T_1 + T_2)]}$$

$$= \frac{E \cdot A_0/T_1 T_2}{s\left[\dfrac{1 + \beta A_0}{T_1 T_2} + s^2 + \dfrac{s(T_1 + T_2)}{T_1 T_2}\right]}$$

$$\therefore \mathscr{L} V_o = \frac{E \cdot A_0/T_1 T_2}{s[s^2 + \alpha s + \gamma]} \qquad \ldots [40]$$

where $\qquad \gamma = \dfrac{1 + \beta A_0}{T_1 T_2}$

and $\qquad \alpha = \dfrac{T_1 + T_2}{T_1 T_2}$

There are three solutions to [40], depending on the values of the constants.

1. If $\gamma > \dfrac{\alpha^2}{4}$, then $s^2 + \alpha s + \gamma = (s+a)^2 + b^2$

   where $\qquad a = \dfrac{\alpha}{2}, \quad b = \sqrt{\gamma - \dfrac{\alpha^2}{4}}.$

   Then $\qquad \mathscr{L} V_0 = E \dfrac{A_0}{T_1 T_2} \cdot \dfrac{1}{s[(s+a)^2 + b^2]}$

   and $\quad V_0 = E \dfrac{A_0}{T_1 T_2} \cdot \dfrac{1}{a^2 + b^2} + \dfrac{1}{b\sqrt{a^2 + b^2}} e^{-at} \sin(bt - \phi)$

   $$\dots [41]$$

2. If $\gamma < \dfrac{\alpha^2}{4}$, then $s^2 + \alpha s + \gamma = (s+a)^2 - b^2$

   and $\quad V_0 = E \cdot \dfrac{A_0}{T_1 T_2} \cdot \dfrac{1}{a^2 - b^2} \cdot \left[1 + \right.$

   $$\left. + \dfrac{(a-b)e^{-(a+b)t} - (a+b)e^{-(a-b)t}}{2b} \right] \qquad \dots [42]$$

   where $\qquad a = \dfrac{\alpha}{2}, \quad b = \sqrt{\dfrac{\alpha^2}{4} - \gamma}$

3. If $\gamma = \dfrac{\alpha^2}{4}$, then $s^2 + \alpha s + \gamma = (s+a)^2$

   where $a = \dfrac{\alpha}{2}$

   and $\qquad V_0 = E \cdot \dfrac{A_0}{T_1 T_2} \cdot \dfrac{1}{a^2} \left[1 - (1 + at)e^{-at} \right] \qquad \dots [43]$

Substituting for a and b, then α and γ in [41], [42] or [43] gives the relevant solution describing the shape of the output response of the closed-loop amplifier to a step input signal.

3.5 Temperature dependence of frequency response

When externally compensated amplifiers are used, there is more control over the performance. However, one or more of the break-points, or the neutralisation of a breakpoint, is often determined by the value of an internal integrated resistor. The resistor can have a temperature dependence which will produce significant effects when amplifiers are operated over wide temperature ranges. It is important to note that external capacitors will have a temperature coefficient effect as well as tolerance errors. General purpose capacitors can have errors of ± 20 per cent or more on tolerance, and temperature effects as large as 0.1 per cent/deg C; however, tolerances of better than ± 1 % are generally available at a premium. Careful attention must be given to these parameters where frequency performance must be optimised. In the case of internally compensated amplifiers, the internal capacitor will also exert tolerance and temperature dependence effects. Table 3.1 gives an indication of the typical temperature coefficients.

Table 3.1 *Capacitor Temperature Coefficients.*

Device	Temperature coefficient ppm/°C
Silicon integrated resistors	up to 2000
Integrated capacitors	± 30 (≡ N.P.O. dielectric)
Polystyrene capacitors	− 50 to − 250
Polycarbonate (metallised, film and foil)	0 to + 200
Mica	−20 to + 70
Polyester (metallised, film and foil)	+ 200 to + 800
Metallised paper	Up to + 2000
Ceramic	± 30 to + 22 − 56 % depending on voltage and temperature range (sometimes to −90 %)
Polyethylene terephthalate (petp)	+ 100 to + 1000
Polypropylene (metallised)	− 300 (typical)

Some more recent devices have integral temperature correction of the frequency compensation, (LM124) thus allowing designs with constant bandwidth over the operating temperature range.

External resistors used in frequency compensation also have temperature dependence, but it is usually easy to obtain resistors with low coefficients for this application, and of less relevance than the effects of resistor errors on gain determination and common-mode rejection.

3.6 Improving closed-loop stability

Stray capacitance associated with the wiring, and more frequently with amplifier input terminals, can introduce breakpoints in the loop response and cause imaginary components to appear in the feedback path. All the foregoing gain analyses and frequency-response determinations have assumed β to be real; in practice it can contain imaginary parts. In Fig. 3.9(a), C_1 represents the stray capacity on the inverting input terminal. The feedback fraction

$$\beta = \frac{R_1}{1+j\omega C_1 R_1} \bigg/ \left[\frac{R_1}{1+j\omega C_1 R_1} + \frac{R_2}{1+j\omega C_2 R_2}\right] \qquad \ldots [44]$$

If $C_1 R_1 = C_2 R_2$, then $\beta = \dfrac{R_1}{R_1+R_2}$ and the effects of the stray capacity are neutralised.

The effective input signal from Fig. 3.9(b) is

$$V_1 = \frac{R_2}{1+j\omega C R_2} V_{\text{in}} \bigg/ \left[R_1 + \frac{R_2}{1+j\omega C R_2}\right]$$

where $C = C_1 + C_2$

$$\therefore V_1 = \frac{R_2}{(R_1+R_2)+j\omega C R_2 R_1} V_{\text{in}} \qquad \ldots [45]$$

In practice, C_1 can cause instability in high-gain situations where $R_2 \gg R_1$, so that small values of C_1 can produce significant phase shifts in the feedback path. For accurate calculations of the effects on gain, the above expression for the effective input signal should be used in the calculations; it can be seen, however, that if $R_2 \gg R_1$, then

$$V_1 \approx \frac{V_{\text{in}}}{1+j\omega C R_1} \qquad \ldots [46]$$

and the frequency at which this becomes significant is usually well outside the band of interest and can be ignored.

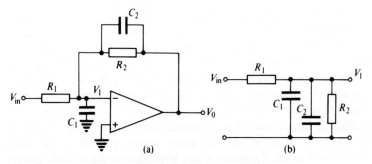

Fig. 3.9 Input terminal capacitance

There are various other frequency-compensation techniques aimed at optimising such parameters as transient response and slew rate without degradation of stability, but these are generally applicable to specific amplifier designs, and information can be obtained from manufacturers' application notes.

Occasionally it is useful to determine the closed-loop bandwidth of internally compensated amplifiers by the use of C_2, in which case $f(3\,\text{dB}) = 1/2\pi C_2 R_2$ and is independent of R_1. The effect of C_2 is to introduce a lead into the feedback path, which does not cause stability problems when the open-loop phase shift is approximately $-90°$; it simply means that the total loop phase shift is $-180°$ rather than the usual $-270°$. This is the reverse of the situation introduced by the presence of C_1, which introduces an extra lag tending to increase the total shift from $-270°$ to $-360°$, the area of danger. The somewhat surprising result for bandwidth determination does not include R_1, and is apparent below. Assume C_1 is zero in Fig. 3.9. Then

$$\frac{V_o}{V_{in}} \approx \frac{R_2}{(1 + j\omega C_2 R_2)R_1} = \frac{R_2}{R_1} \cdot \frac{1}{1 + j\omega C_2 R_2}$$

$$\left| \frac{V_o}{V_{in}} \right| = \frac{R_2}{R_1} \cdot \frac{1}{\sqrt{1 + \omega^2 C_2{}^2 R_2{}^2}} \qquad \dots [47]$$

at $-3\,\text{dB}$,
$$\left| \frac{V_o}{V_{in}} \right| = \frac{R_2}{R_1} \cdot \frac{1}{\sqrt{2}}$$

$$\therefore \omega^2 C_2{}^2 R_2{}^2 = 1 \qquad \text{and} \qquad f(-3\,\text{dB}) = \frac{1}{2\pi C_2 R_2}$$

3.7 Common-mode rejection

This was defined in Section 2.1.3, and is usually quoted at d.c. or a low frequency. The common-mode rejection ratio (CMRR), which is of great advantage in noisy environments (since pickup can be arranged to be almost equal to both input terminals in a properly designed balanced system) has potentially the greatest advantage at 50 Hz line frequency and above in the rejection of unwanted signals. Consequently a more useful parameter would be the ratio at defined frequencies, a figure often not specified. For most amplifiers the ratio begins to reduce at low frequencies, and so the d.c. value can give a misleading impression of performance when low-level signals are to be amplified in an electrically noisy situation. This is especially important in control systems where large electromagnetically generated transients can occur, as with relay and switching-type circuits.

The effects can of course be minimised by careful attention to wiring and shielding, and as a general rule all electrical noise generation should be minimised at the source by suppression techniques in the first instance. A useful reference work has been produced by Morrison.[10] Ultimately the amplifier itself then provides further noise rejection; however, it is clear that the actual ratio as defined must reduce quickly for increasing frequency, as the input equivalent error is a function of the open-loop gain, and the latter rolls off at low frequencies for internally compensated amplifiers. It might be expected that the device internal imbalance which gives rise to the error signal would also reduce with frequency, thus maintaining a constant ratio, but this is usually not so, since the frequency-dependent error is a result of internal reactive imbalance. This problem of mismatch between stray capacities is nowhere more relevant than at the device input terminals, and it is well to remember that the high common-mode rejection performance of the device is invariably lost as soon as external connections are made. The effects of external resistor mismatch are discussed in Chapter 4, but here we will assume that the resistors are perfect. An expression has been derived in Appendix 4 for the effective input error signal (see section 1.8.3) resulting from capacitive mismatch at the input terminals and generated by a common mode voltage:

$$V_{error} = V_{cm}R_2 \cdot \frac{\omega C R_1 R_2 (n-1)}{\sqrt{[(R_1+R_2)^2 - n\omega^2 C^2 R_1{}^2 R_2{}^2]^2 + (R_1+R_2)^2 \omega^2 C^2 R_1{}^2 R_2{}^2 (1+n)^2}} \quad \dots [48]$$

where V_{cm} = common-mode input
R_1 = input resistor
R_2 = feedback resistor
$\omega = 2\pi f$
n = ratio of C_{in_1} to C_{in_2}

A typical situation could be as follows:

$R_1 = 10 \text{ k}\Omega$
$R_2 = 100 \text{ k}\Omega$
$C = 10 \text{ pF}$
$n = 2$ (so that the other capacity = 20 pF)
$f = 1 \text{ kHz}$

Substituting in the above equation for a 10 V common-mode signal gives $V_{error} = 5.2$ mV. Now this may well be a large proportion of the actual desired input signal, and this at a modest 1 kHz and input capacity of 10 pF which is not at all unrealistic. It is obvious that if a.c. common-mode rejection is to be maintained, attention to external connections is imperative.

The external problems will add to the error signals, but the actual amplifier performance also deteriorates with frequency due to internal effects. Typical curves are reproduced in Fig. 3.10. It is important to remember that common-mode rejection (even at d.c.) tends to be a non-linear function of the input common-mode operating voltage (i.e. the bias on both input terminals), and so the expected performance may only be achieved at specified common-mode levels. Figure 3.10 shows the variation of CMRR with frequency for some common amplifiers. It is interesting to note that in the case of (b) where the amplifier loop has been closed, the ratio remains substantially constant with frequency. With some amplifiers the common mode rejection is affected by adjustments to the input offset voltage, so optimisation must involve a consideration of this effect.

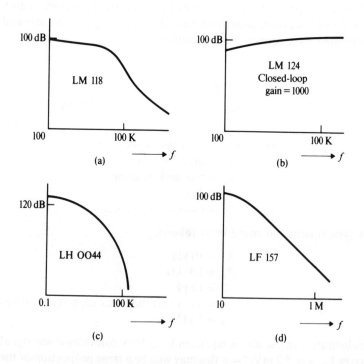

Fig. 3.10 Examples of common-mode rejection frequency dependence

3.8 Slew rate limiting[11]

The maximum rate of change of output voltage for compensated amplifiers is usually defined by the ability of the device(s) to which the frequency-compensating capacitor is connected to drive current into this capacitor. Since $q = CV$ and $i = dq/dt = C(dv/dt)$, i is directly proportional to C and the rate of change of voltage. During the evolution of bipolar op amps, various methods have been used to improve the current-driving capability of the frequency-compensating stages. Another, perhaps less important, limiting mechanism is the ability of the output stage to drive current into capacitive loads.

If $V_o = V\sin 2\pi f_m t$, then

$$\frac{dV_o}{dt} = 2\pi f_m V\cos 2\pi f_m t$$

and the 'slew rate' or maximum rate of change of output voltage is $S = 2\pi f_m V$ where V is the peak output voltage swing at f_m, the 'full-power bandwidth' as it is sometimes called. It can be seen that as the frequency is increased, the output voltage swing can be decreased without slew rate limiting, so that in practice an amplifier can provide low-level undistorted outputs at much higher frequencies than would at first seem possible. Various frequency compensating techniques are recommended by I.C. manufacturers to improve the slew rate, and reference should be made to individual data sheets. Invariably there is a trade-off between slew rate improvement and some other parameter such as noise or stability margin. Feedforward is one possible technique which is described in Chapter 6.

Since voltage followers require the low-frequency compensation roll-off (so that phase shift is limited at unity closed-loop gain) it is often difficult to maximise slew rate; recourse to the manufacturer's recommendations on particular devices is desirable. Slew rates of early op amps were in the order of 0.1 V/μs, but later devices are capable of 100 V/μs and more. The small-signal response of an amplifier can be more clearly determined by defining small signal as 'any signal which does not cause slew rate limiting' and this is often the implicit statement when describing small signals. Alternatively, f_m in the above expressions can be defined as 'the maximum frequency at peak output, below which slew rate distortion does not occur'. One is prompted to ask 'what is actually meant by distortion?' in this context, since it is a relative term. In fact it is rarely qualified, and photographs of oscilloscope traces are used to illustrate the observed effect on waveform distortion. We have seen that rise time of a simple single-pole circuit is

$$t_r = \frac{0.35}{f_0} \text{ where } f_0 \text{ is the } -3 \text{ dB point.}$$

When the output step voltage V_S divided by the rise time is greater than the slewing rate S, the output will become a ramp with a slope of S and a rise time of $t_r' = \dfrac{V_S}{S}$, irrespective of the speed of the input waveform. Hence

$$S \approx \frac{V_S f_0}{0.35}$$

and this will give a quick check on the slew rate, since it is easy to measure. What then, does this really mean for a sinusoidal input? In effect, we are injecting a sinusoid of $V = A\sin \omega t$ into a circuit with a maximum rate of change of output dV_o/dt of S. Now the maximum slope of the sinusoid occurs at the zero crossing point, so that if the slope of the sinusoid at zero is greater than S, the amplifier output will not be able to follow. One might expect the result to look like Fig. 3.11. The tendency is for the output waveform to become triangular in shape, but often asymmetric. Hence distortion components are introduced into the fundamental waveform, and if it is triangular in nature is likely to include largely odd harmonic content. In theory it

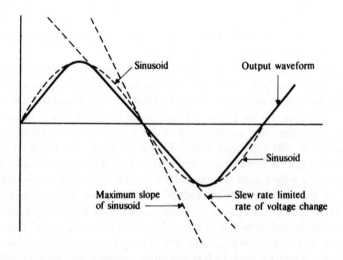

Fig. 3.11 The effects of slew rate limiting on a sinusoid

should be possible to measure the harmonic distortion and correlate this with the sinusoid frequency and the slew rate, although the author has not tried this in practice.

There are a few devices for which a curve is given showing the maximum obtainable output voltage swing variation with frequency for distortion of less than 1 per cent (e.g. LM 148).

During the development of bipolar ICs, various improvements have been incorporated in the circuit so that the frequency-compensation capacitor can be driven at higher dV/dt, including special current-drive stages. It is a fact, however, that field effect transistors (FETs) have inherently higher slew capability than bipolar devices, so it is a logical step to use both FET and bipolar devices on the same chip as the technology of manufacture progresses. As a fundamental comparison, for either device the slew rate is proportional to the maximum output current I_0 divided by the mutual conductance (transconductance) gm. So that for a bipolar device, $\dfrac{I_0}{gm}$

$$\propto \frac{KT}{q} = 26 \text{ mV, and for an FET, } \frac{I_0}{gm} \approx \frac{V_p}{2} \approx 500 \text{ mV}$$

where K = Boltzman's constant q = electron charge
T = absolute temperature V_p = pinchoff voltage

Later circuits using FET stages to provide high current drive, and bipolar outputs for their low source resistances, have improved slew capability. A comparison of some typical circuits is given in Table 3.2, based on open-loop curves.

The values have been calculated from $S = 2\pi f_m V$, where f_m is the maximum frequency at which the peak output voltage V is available, taken from data curves. Sometimes tabulated data gives S for unity closed-loop gain, but this is bound to be a lower value in most cases since the single-pole roll-off low-frequency compensation is used for the unity gain configuration. It should be noted that achievable slew rate sometimes differs slightly from the calculated values quoted here.

3.9 Capacitive loading

A common use for op amps is for signal preamplification to drive long lines. Cables, especially screened cables, have high capacities per unit length, and so a long line can produce high equivalent capacitive loading on the output terminal. It is not uncommon for a perfectly stable amplifier to oscillate when it is connected to a long screened

Table 3.2 *Slew rate of some available operational amplifiers.*

Type	Calculated slew rate (V/μS) (typical)	
LM124	0.42	
LM107	0.44	
μA741	0.88	
LM4250	1.8	
LM149	2	(Closed-loop gain × 5)
RC4136	2.2	
LM3900	2.6	(Average for positive and negative swings)
HA4602	3.8	
CA3140	7.9	
μA748	8.5	
LF156	15.7	
TDA1034	16	
μA709	21	
μA725	30	
SE531	35	(Closed loop gain × 100)
SE538	55	
μA715	70	(Closed loop gain × 100)
BB3507J	80	
LM118	90	
AD505	120	
TP1434	120	(quoted, not calculated)
AM405–2	126	
ICL8017	163	
BB3554	1 000	(quoted)

output line. It was shown in Section 2.3.3 that the low-frequency output resistance of an amplifier with feedback is

$$R_C = \frac{R_0}{1 + \beta A}$$

where R_0 is the amplifier output resistance, and A the low-frequency open-loop gain. It has been convenient to let A_0 = d.c. gain and A = gain at any frequency. For a simply compensated amplifier

$$A = \frac{A_0}{1 + j(f/f_0)}$$

where f_0 is the open-loop -3 dB frequency. The effective output resistance R_C must change with the open-loop gain, since

$$R_C = R_0 \cdot \frac{1}{1 + \beta A_0/[1 + j(f/f_0)]} = \frac{1 + j(f/f_0)}{(1 + \beta A_0) + j(f/f_0)} \times R_0$$

$$R_C = \frac{[1 + \beta A_0 + (f/f_0)^2] + j\beta A_0(f/f_0)}{(1 + \beta A_0)^2 + (f/f_0)^2} \times R_0 \qquad \ldots [49]$$

Calculating the magnitude yields

$$R_C = R_0 \sqrt{\frac{1 + (f/f_0)^2}{(1 + \beta A_0)^2 + (f/f_0)^2}} \approx \frac{R_0(f/f_0)}{\sqrt{(1 + \beta A_0)^2 + (f/f_0)^2}} \quad \dots [50]$$

Taking as an example $A_0 = 100\,000$, $\beta = 1/100$ and $(f/f_0) = 1000$, where $f_0 = 10\,\text{Hz}$ (i.e. $f = 10\,\text{kHz}$), the effective output resistance is $R_0/\sqrt{2}$. Now this compares with the low-frequency equivalent output resistance of

$$R_C = \frac{R_0}{1 + \beta A_0} = 0.001\,R_0.$$

As the frequency increases, the effective output resistance tends to the open-loop value. The load capacity C_L in Fig. 3.12 is thus driven from an increasing source resistance as the freqeuncy goes up, and produces another phase lag within the loop. At high frequencies, the total loop phase lag is likely to be more than $-270°$ (i.e. an inversion $\equiv -180°$, + compensation lag, + stray lag), to which is added the effects of R_C and C_L. This is the situation appertaining when a compensated amplifier oscillates when loaded with a long line. In order to avoid this difficulty, we must isolate the effects of C_L on the loop; this is most easily accomplished by inserting a low-value resistor directly in series with the amplifier output, or by feedback decoupling as in Fig. 3.9 ($C_1 = 0$). Inserting a resistor has the advantage that the absolute value of C_L tends to be less important, but the low-frequency output resistance is degraded to the value of the resistor inserted. A typical value would be $10\,\Omega$. The circuit then becomes equivalent to Fig. 3.12(c) with R_C tending to R_0 at higher frequencies (assuming R_2 does not load the output). Let V_f = voltage at the feedback point, then

$$\frac{V_f}{V_o} = \frac{r + (1/j\omega C_L)}{R_o + r + (1/j\omega C_L)} = \frac{(1 + j\omega C_L r)[1 - j\omega C_L(r + R_o)]}{1 + \omega^2 C_L^2 (R_o + r)^2}$$

$$= \frac{1 + \omega^2 C_L^2 r (R_o + r) - j\omega C_L R_o}{1 + \omega^2 C_L^2 (R_o + r)^2}$$

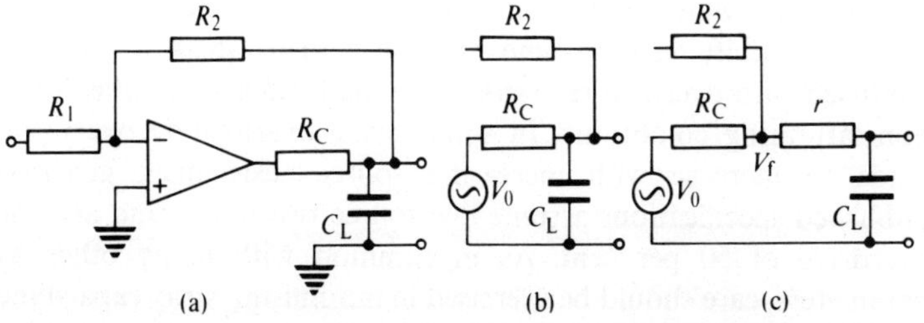

Fig. 3.12 Isolation of capacitive output loading

The phase shift to V_f is

$$\phi = -\tan^{-1}\left[\frac{\omega C_L R_o}{1 + \omega^2 C_L{}^2 r(R_o + r)}\right] \qquad \ldots [51]$$

Suppose then that we wish to limit the additional phase lag due to C_L to $30°$ at 1 MHz, the open-loop unity gain frequency. Then $\tan\phi = 0.58$, and let us assume that $C_L = 10\,000$ pF and $R_o = 100\,\Omega$. Then solving for

$$\tan 30° = 0.58 = \frac{\omega C_L R_o}{1 + \omega^2 C_L{}^2 r(r + R_o)}$$

gives $r^2 + 100r - 2491 = 0$

and hence $r \approx 21\,\Omega$

It should be remembered that the low-frequency output resistance is now also approximately $21\,\Omega$, so the effects on accuracy when the output drives into a specified load will need to be considered.

3.10 Overload recovery

This is the time taken for the amplifier to recover its initial small-signal operating conditions after saturation. This is a complex parameter, and careful attention must be paid to the conditions for which it is specified. For example, it is possible for the output-stage transistors to be saturated without any of the other stages being overdriven, or for the input to be overdriven with a large signal; the recovery time in these two cases may be quite different. Alternatively, the feedback signal itself, and not the source may drive the input to saturation. Frequently the principal factor determining the recovery is the nature and driving source of the frequency-compensation circuits; amplifiers with a high roll-off rate tend to recover more quickly than 6 dB/8^{ve} devices. High current output amplifiers may draw pulses of current from the power supply which can depress the rail voltages, which in turn affects the recovery time. Some manufacturers specify 'settling time' (section 2.1.16), which is not to be confused with overload recovery, since the former is specified under non-saturating conditions. In some cases, overload recovery time tends to increase with increasing source resistances; generally, published specifications assume low source resistances and an input overdrive of 50 per cent. As in common with many other a.c. parameters, care should be exercised in minimising stray capacitance in the feedback connections.

3.11 Power supply rejection ratio (PSRR)

As with CMRR, the PSRR also falls rapidly with frequency, and again tends to roll off from similar frequencies and at similar rates to the open-loop response. The shape of the curve may well be different for the positive and negative rails; wider-bandwidth amplifiers tend to have improved rejection at higher frequencies. Where switching or oscillator circuits are likely to be operating from the same power supply as an amplifier, considerable attention should be given to the mechanism of supply noise generation (and of course where amplifiers are driving high load currents as with audio applications). For example, a current switched in one part of the circuit can generate high-level fast voltage transients due to the presence of line inductance between the circuit and the power source, thus presenting a high impedance in the line at high frequencies. For example, the PSRR at 100 kHz may be quoted as 20 dB (assume that the figure given applies at d.c. if no supporting frequency data is provided). This is a ratio of only 10, and means that a 1 V rail transient appears as an input signal of 0.1 V – and could easily saturate the amplifier output in small-signal applications. The result is not simply a 100 kHz component in the signal, but perhaps a much longer component resulting from the overload recovery time. The usual method of minimising these problems, and also possible feedback from the amplifier output via the rail in high-frequency applications, is to insert a simple filter in the rail connections as in Fig. 3.13. Note that R is chosen so that the voltage drop due to the amplifier supply current is not sufficient to cause an unacceptable change in amplifier performance (i.e. not only gain, but such parameters as common-mode range and output swing required). C_1 may then be any suitable high-frequency capacitor (e.g. ceramic or polystyrene) so that adequate rejection is obtained at the frequency of interest. Low-frequency rejection may be improved by including C_2, electrolytic capacitors; since they tend to be inductive it is not advisable to use these alone, except perhaps tantalum-bead types. For the simple R–C network, the relative response $\delta V_{CC}/\delta V_A$ is $\sqrt{1 + (f/f_0)^2}$, so that the overall rejection at the frequency f of interest is $= \mathrm{PSRR}_f + 20\log_{10}\sqrt{1 + (f/f_0)^2}$, where $f_0 = \dfrac{1}{\omega CR}$.

Another source of power-supply noise which can cause difficulty, especially in bridge-sensor connections, or where CMRR must be maximised, is transient feedthrough from the power supply transformer. The mechanism of noise generation (other than direct mains-borne interference) can be understood by reference to Fig. 3.14. The capacitor C_S represents inter-winding stray capacitance, R_L the load

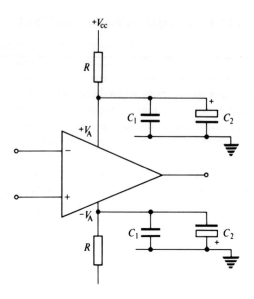

Fig. 3.13 Line filtering

and C the reservoir. Current will only flow into the load as long as V is more than one diode drop greater than V_o. As soon as the load current stops, an inductive source (the transformer) has been interrupted, and there is the possibility of high-voltage transient generation across the transformer secondary. The tendency is for this transient to introduce a current flow through C_S and the source, and the loop is then completed by any connection R, between V_{01} or V_{02} and ground. Thus any monitoring equipment connected in this way, and any

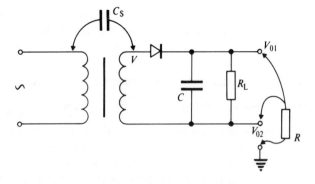

Fig. 3.14 Noise coupling through power supply transformer inter-winding capacity

differential input amplifier connected to a bridge source where the amplifier common line is referred to ground, can experience noise spikes. There is a choice of methods to minimise this type of problem:—

1. Centre tapped to ground transformer secondary.
2. Toroidal transformer (low C_S).
3. Interwinding earthed screen, and equipotential secondary screen.
4. Possible speed degradation by connection of a C–R series combination across the diode.

Several of the above may well be desirable features in any low-level circuit, in order to eliminate power supply electromagnetic radiation interference and mains-borne transients.

3.12 Active rectifiers

Figure 3.15 (a) and (b) are the well-known basic circuits for active half-wave and full-wave rectification respectively. Let us consider the operation of the simple half-wave circuit, since the

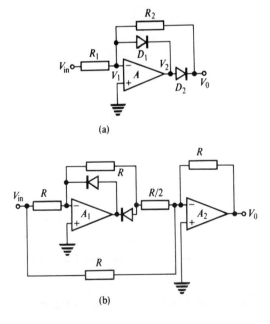

(a)

(b)

Fig. 3.15 Half-wave (a) and full-wave (b) active rectifier circuits

same frequency-limitation arguments apply to both circuits. For negative-going input signals, D_1 is reverse biased, and

$$V_o = V_2 - V_d \qquad \text{where } V_d = \text{diode voltage drop}$$
$$V_2 = AV_1$$
$$V_1 = \frac{R_2 V_{in}}{R_1 + R_2} - \frac{R_1}{R_1 + R_2} V_o$$

From the above,

$$V_o = \frac{AR_2}{R_1 + R_2} V_{in} - \frac{AV_o R_1}{R_1 + R_2} - V_d$$

$$\therefore V_o \left(1 + \frac{AR_1}{R_1 + R_2}\right) = \frac{AV_{in} R_2}{R_1 + R_2} - V_d$$

$$V_o = V_{in} \frac{AR_2}{R_1(1 + A) + R_2} - \frac{V_d}{1 + AR_1/(R_1 + R_2)} \qquad \dots [52]$$

If A is very large,

$$V_o = V_{in} \frac{R_2}{R_1} - \frac{V_d}{1 + \beta A} \qquad \dots [53]$$

It can be seen then, that the diode voltage drop has only a very small effect on the output voltage, being reduced by the loop gain. Consequently, the temperature dependence of V_d, about $-2\,\text{mV}/°\text{C}$, is also reduced by $1/(1 + \beta A)$ and good stability with temperature is obtained. For a positive-going input, D_1 conducts and clamps the output to the amplifier input, thus preventing the amplifier from going into negative saturation. The value of A for the simply compensated amplifier is expressed by

$$A = \frac{A_0}{1 + j(f/f_0)} \quad \text{where } f_0 = -3\,\text{dB open-loop frequency.}$$

Consequently, [52] for a.c. applications becomes

$$V_o = \frac{V_{in} R_2 A_0}{(R_1 + R_2)[1 + j(f/f_0)] + A_0 R_1} - \frac{V_d}{1 + \beta A_0/[1 + j(f/f_0)]} \qquad \dots [54]$$

Calculating the magnitudes as in [36] and [50] gives

$$V_o = \frac{AV_{in} R_2}{\sqrt{[R_2 + R_1(1 + A_0)]^2 + (R_1 + R_2)^2 (f/f_0)^2}}$$
$$- V_d \sqrt{\frac{1 + (f/f_0)^2}{(1 + \beta A_0)^2 + (f/f_0)^2}} \qquad \dots [55]$$

Although this expression explains the closed-loop performance of the rectifier, there is an important fact which has not yet been considered. Assume that the amplifier has zero offset voltage, so that 0 V input truly produces 0 V out. In this condition the amplifier is in open-loop operation, since D_2 cannot be conducting. During the time that V_2 is changing from 0 V to $+V_d$, the amplifier is open-loop, and for good a.c. rectification this time must be as short as the time taken for the input waveform to change from 0 to $+V_d$. Now for a simply compensated amplifier, the open-loop response is likely to have a -3 dB frequency of about 10 Hz or less, which tends to indicate a low slew rate. This is in direct contradiction to the required performance, since we wish the output to slew as fast as possible from 0 to $+V_d$. In most cases then, the need for stable unity closed-loop gain is counter to the need for high slew rate (or high full-power bandwidth). Consequently, for good a.c. rectification we require an amplifier which is compensated for unity closed-loop gain but has high slew; a figure of merit can thus be derived by comparing the maximum full-power bandwidth available when unity closed-loop gain stability is incorporated. It would be expected that the maximum frequency at which good rectification can be achieved is less than $f = S/2\pi V$; for example, if $S = 1$ V/μs, $V = 10$, then $f = 15.9$ kHz. However, even at this frequency, the diode drop V_d of 0.6 V would take the amplifier 0.6 μs to slew through. At 15.9 kHz, $T/2 = 31$ μs, so that the amplifier is effectively open-loop (and thus not defined) for 2 per cent of the period. For a 741 type (Table 3.2), it is unlikely that good rectification beyond 10 kHz is possible (the slew rate must be applicable at unity closed-loop gain since the amplifier is internally compensated). Table 3.3 tabulates the maximum frequency at which the peak output voltage is available, using the necessary frequency compensation for unity closed-loop gain stability, as a figure of merit for some common devices. For further information about circuit improvements, refer to references[12]. There are a number of hybrid circuits which use two or more chips packaged in IC-type cases, but the discussion above has been limited to single-chip ICs. The dual-chip devices may well provide improved performance, but generally cost more than the single-chip types.

3.13 Output current drive limiting

We have seen (section 3.8) the effects of slew rate limiting, and considered the results of capacitive loading. The problem of output current drive can also cause an apparent decrease of slewing speed with resistive loads. As the load resistance is decreased, the output

Table 3.3 *Selection of operational amplifiers for active rectifiers*

Type	Max frequency for peak output when compensated for unity gain (kHz)
μA725	0.05
LM108	1.5
μA709	3
LM124	5
μA741	10
μA748	10
LM3900	20
RC4136	25
SE531	50
HA4602	50
TDA1034	90
μA715	100
CA3140	100
TLO81	150
LF156	200
OP-16	320
LM118	1 200
AM-405-2	2 000
ICL8017	2 000
AD505	2 500

voltage swing also decreases even at low frequencies, due to internal voltage drop across limiting resistors and increasing output transistor saturation voltages. The effect is usually non-linear, so that 5 mA may be available at 10 V or 2 mA at 12 V, for example. As a result of this, if the load resistance is low the non-linear output can appear to be the same as slew rate limiting, even though the normal mechanism of slew limit (i.e. current driving of the internal compensation capacitor) is not acting.

4

Effects of external components

4.1 Resistors

These are by far the most important components associated with analog applications, and it is essential that all designers should be familiar with the various types available and their performance. The performance of accurate low-level integrated circuits is often degraded by the inadequate selection of the components associated with them. Variable resistors and trimming potentiometers can never have the same overall performance as a fixed resistor, so the use of potentiometers must be carefully considered. For example, in areas of high vibration a potentiometer should be selected with specific regard to this parameter; if available components do not meet the necessary specification it is often better to use select-on-test fixed resistors, although the cost/performance compromise is an essential factor in a production environment. Table 4.1 has been drawn up to give an indication of the performance capabilities of fixed resistor types, although it should not be taken as an ultimate reference; many manufacturers will supply to specific requirements where quantities merit or costs are met. A typical parameter is selection tolerance; standard production tolerances are normally stocked, but much closer tolerances may be available on request. Frequently, of course, this is simply transferring the onus of selection to the supplier from the user, but the speed with which the supplier can do the testing often produces a more economic solution to the user. Bear in mind that if you, the user, order a batch of ± 2 per cent tolerance resistors, you should not necessarily expect to receive a standard distribution around the nominal value with cut-offs at $+$ and $- 2$ per cent. It may just be that the supplier has already selected out all the resistors of nominal ± 0.1 per cent value for another customer. Under these circumstances you would not get any components within that range. In addition, you will be interested not only in the electrical

Table 4.1 Typical resistor availability

Resistor type	Standard selection tolerance (%)	Typical resistance range (ohms)	Temperature coefficient of resistance (ppm/°C)	Stability 1 year shelf (%)	Stability 2000 hours 70°C; full rating (%)	Power range (W)	Operating temperature range (°C)	Typical noise (µV/V)
Metal glaze (thick film)	1–5	1R–33M	±100 to ±300	0.1	1	$\frac{1}{4}$–6	−55 + 150	0.1
Wirewound	0.005–1	0R1–10M	±2 to ±200	0.005	0.01	$\frac{1}{8}$–100	−55 + 155	Johnson noise
Moulded carbon	2–10	1R–10M	±1200	5	15	$\frac{1}{8}$–2	−40 + 105	$2 + \log_{10} \dfrac{R}{10^3}$
Carbon film	1–10	1R–120M	−200 to −1200 depending on R	2	4	$\frac{1}{16}$–5	−40 + 125	1
Metal oxide	1–5	1R–1M	±100 to ±300	0.1	1	$\frac{1}{8}$–2	−55 + 150	0.4/decade bandwidth at 1M. Increases with R.
Metal film	0.05–5	1R–30M	±10 to ±150	0.1	0.2	$\frac{1}{16}$–2$\frac{1}{4}$	−55 + 150	Johnson
Precision metal film	0.001–1	2R5–5M	±1 to ±10	0.0025	0.0025	$\frac{1}{16}$–$\frac{1}{2}$	−55 + 100	Johnson noise
Cermet (pots)	5–20	10R–2M	±50 to ±200			$\frac{1}{2}$–2	−55 + 150	Johnson noise
Conductive plastic (pots)	10 typ	300R–40K	±100 to ±300			1–8	−65 + 130	Johnson noise +
Hybrid plastic + wire (pots)	10	5K	±80			1	−65 + 130	Wiper noise

Johnson noise = $e_{\text{rms}} = \sqrt{4kTR\Delta f}$

K = Boltzman's const.

T = Absolute temp.

Δf = Bandwidth

performance, but also the physical characteristics such as body construction (e.g. lacquered, dipped, moulded, ceramic, etc.), hotspot temperatures, resistance to chemical attack, etc. Table 4.1, which is only a rough guide to the availability of resistors, shows the very large range of performance parameters. After selection tolerance, probably the most important features of interest to the designers of op amp circuits are stability and temperature coefficient. Let us consider an example. Assume a temperature coefficient of ± 200 ppm/°C = $\pm 0.02\,\%$/°C. An amplifier with a fixed gain determined by such resistors will change gain with temperature. Consider the two op amp configurations in Fig. 4.1(a) and (b).

From (a), $G = \dfrac{R_2}{R_1}$. Then

$$\frac{dG}{G} = \frac{\partial G}{\partial R_2}\frac{dR_2}{G} + \frac{\partial G}{\partial R_1}\frac{dR_1}{G} \qquad \ldots [56]$$

$$\therefore \frac{dG}{G} = \frac{1}{R_1}\frac{dR_2}{G} - \frac{R_2}{R_1{}^2}\frac{dR_1}{G} \qquad \ldots [57]$$

Percentage change of gain is $100 \times \dfrac{dG}{G}$

$$= 100 \times \frac{dR_2}{R_2} - \frac{dR_1}{R_1}.$$

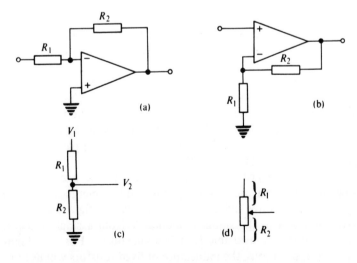

Fig. 4.1 Gain dependence on resistor temperature coefficient

If the temperature coefficient of both resistors is identical, then the gain change will be zero. However, if we take the specification limit of $+0.02\%/°C$ for R_2 and $-0.02\%/°C$ for R_1, the overall gain change could be $0.04\%/°C$, which over a $100°C$ operating range represents a 4 per cent gain change!

Consider also Fig. 4.1(b):

$$G = \frac{R_1 + R_2}{R_1}$$

$$\frac{dG}{G} = \frac{1}{R_1}\frac{dR_2}{G} - \frac{R_2}{R_1{}^2}\frac{dR_1}{G}$$

$$= \frac{R_2}{R_1 + R_2}\frac{dR_2}{R_2} - \frac{R_2}{R_1 + R_2}\frac{dR_1}{R_1} \qquad \ldots [58]$$

The gain change in this case depends on the relative values of R_2 and R_1. The larger the gain, the nearer the change approaches that for the inverting configuration, as $\frac{R_2}{R_1 + R_2} \to 1$. For lower gains, the maximum temperature dependence decreases from that of the inverting configuration.

A similar argument applies to the attenuator in Fig. 4.1(c), as $V_2 = \frac{R_2}{R_1 + R_2}V_1$. Then

$$\frac{dV_2}{V_2} = \frac{R_1 + R_2}{R_1}\frac{dR_2}{R_2} - \frac{R_1 + R_2}{R_1}\frac{dR_1}{R_1} \qquad \ldots [59]$$

and the resulting temperature-dependent output depends on the coefficients of R_1 and R_2 and the relative values. Note that if the temperature coefficients are equal, the output remains constant; hence in some applications matched resistors can be used to advantage. A similar situation is inherent with the use of a potentiometer as a divider, since it might reasonably be expected that the temperature coefficient of resistance will be equal for both halves of the potentiometer (Fig. 4.1 (d)). Table 4.2 gives some examples of resistor coding from British Standard BS1852.

Improved stability can often be obtained by operating resistors well below their nominal power-rating capability; sometimes they are 'triple rated', i.e., three different power dissipation figures are given corresponding with the stability obtainable from a single resistor. Detailed curves are available from manufacturers. For very high-frequency applications, the inductance of fixed resistors will need to be considered. Wire-wound resistors are often worse than solid types,

Table 4.2 *Resistor marking*

Selection tolerance code	Resistor values
F = ± 1%	0.47 Ω = R47
G = ± 2%	3.9 Ω = 3R9
J = ± 5%	10 kΩ = 10K
K = ± 10%	3.3 MΩ = 3M3
e.g. 4M7M = 4.7 MΩ ± 20%	

unless they are specifically described as 'non-inductively wound'. Resistor self-capacitance is rarely a problem.

4.2 External d.c. balance

The external terminals provided on some amplifiers are very useful for removing the effects of initial offset voltage, but their range is usually limited to ± 2 or 3 times the maximum initial value, i.e. about ± 15 mV. In many applications it is desirable to remove the initial errors generated by transducers or other input sources, by applying an offset to the amplifier in the opposite sense. In these cases the control terminals provided are not adequate, and alternative techniques must be used. It is therefore convenient to separate a discussion of balancing into two parts, those using extra circuit techniques and those using the amplifier terminals directly. There is an important point to remember whenever the design of an amplifier is considered: always make sure that any initial zero errors are eliminated before applying a variable gain control, otherwise a change of gain will apply to the initial error voltage and produce a shift of the reference datum.

4.2.1 External circuits

For either inverting or non-inverting configurations, balancing signals can be applied to either input, but the choice will depend on specific applications: some schemes have disadvantages compared with others.

One possibility is shown in Fig. 4.2(a). Remember that V_x is a virtual earth point, so the apparent resistance to external voltages is zero at this point.

$$V_o = \frac{-R_2}{R_1} V_{in} - \frac{R_2}{R_3 + R_4} V_1 \qquad \ldots [60]$$

where $R_4 = \frac{R_5 R_6}{R_5 + R_6}$ and $V_1 = \frac{R_5}{R_5 + R_6} V_{cc}$

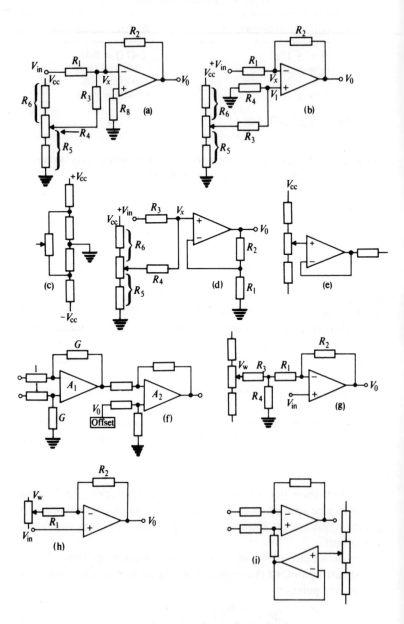

Fig. 4.2 Offset voltage balancing techniques

$$\therefore V_o = \frac{-R_2}{R_1} V_{in} - \frac{R_2}{R_3 + R_5R_6/(R_5 + R_6)} \cdot \frac{R_5}{R_5 + R_6} V_{cc}$$

For the balanced condition where $V_o = 0$,

$$\frac{V_{in}}{R_1} = - V_{cc} \frac{R_5}{R_3(R_5 + R_6) + R_5R_6} \qquad \cdots [61]$$

From [61] it is clear that V_{cc} must be negative, and a rather complicated relationship results. However, by making the resistance of the potentiometer chain small compared with R_3 (N.B. consider the supply current drawn) the loading by R_3 is minimised, and

$$V_o = \frac{-R_2}{R_1} V_{in} - \frac{R_2}{R_3} V_w \qquad \text{where} \quad V_w = \text{potentiometer wiper voltage}$$

for balance, $V_w = \dfrac{-R_3}{R_1} V_{in}$.

In all situations where the loading of R_3 is significant, the offset control will be non-linear. In order to minimise the offset generated by input bias current, R_8 is made equal to the source resistance at the inverting input terminal, in this case approximately R_1, R_2 and R_3 in parallel.

In Fig. 4.2(b), the objective is to make $V_1 = V_x$ (V_x is the effective input voltage – see 1.8.3) to null the effects of $+ V_{in}$. V_{cc} must therefore be positive. For balance,

$$V_1 = \frac{R_2}{R_1 + R_2} V_{in} = \frac{R_4}{R_3 + R_4} V_2 \qquad \cdots [62]$$

where $\qquad V_2 = \dfrac{R_7}{R_6 + R_7} V_{cc} \quad \text{and} \quad R_7 = \dfrac{(R_3 + R_4)R_5}{R_3 + R_4 + R_5}$

Again, if $R_3 = R_2$ and $R_4 = R_1$, and the loading of $R_3 + R_4$ on the potentiometer chain can be ignored, the wiper voltage must equal the input voltage for balance. For minimisation of bias current generated offset voltage, the source resistances in each input terminal should be made equal, i.e.,

$$\frac{R_1R_2}{R_1 + R_2} \approx \frac{R_3R_4}{R_3 + R_4}$$

The loading of R_3 and R_4 on the potentiometer will again produce a non-linear control. Sometimes a more convenient arrangement for defining small positive and negative reference levels is used, as shown in Fig. 4.2 (c). R_4 may be omitted completely if there are no common-mode problems; in this case, since the amplifier input resistance is usually very high, the wiper voltage will be equal to V_x, there will be negligible loading, and the control will be linear.

In Fig. 4.2(d), R_3 is taken to include the source resistance (R_3 may be required for input protection or terminal resistance balancing). With this circuit it is difficult to apply an offset correction to the inverting input, since any loading will affect the feedback fraction and thus the gain. The gain here is

$$V_o = \frac{R_1 + R_2}{R_1} \times V_x \quad \text{and} \quad V_x = \frac{R_4 + R_7}{R_4 + R_3 + R_7} V_{in} \quad \text{where}$$

$$R_7 = \frac{R_5 R_6}{R_5 + R_6}$$

so that the gain to V_{in} varies with adjustment of the pot if R_7 is significant compared with R_4. For a balance, V_{cc} must be negative, and

$$\frac{V_{in}}{R_3} = \frac{V_{cc} R_5}{R_6(R_4 + R_5) + R_4 R_5}, \quad \text{or if } R_4 \gg R_7, \quad \frac{V_{in}}{R_3} = \frac{V_w}{R_4}.$$

Differential input configurations can be accommodated by arrangement (b), connecting the common return of R_4 to the other input source. It should be noted that the resistors in series with the potentiometer may be removed in some applications; the criterion here is range of control required compatible with the desired resolution. Even if cermet flatpots are used with their theoretically infinite resolution, the sheer physical difficulty of making a small enough adjustment (not to say overcoming the lead screw backlash) demands a reduction of range. Sometimes multiturn helical potentiometers may be used, but these are expensive. Another alternative is to use two potentiometers arranged to give coarse and fine controls; in this case, the stability of the coarse control will be most important (e.g. the effects of shock and vibration). As with the use of fixed resistors, the effects of temperature coefficient of resistance will need to be considered over the operating temperature range.

With the advent of single packages containing several amplifiers, the follower circuit, Fig. 4.2(e), is often very convenient. It can be used in any of the above configurations, and offers the advantages of linear control (easily defined voltage levels), low output resistance, and hence ease of balancing amplifier input terminal resistances. It is also very easy to use other amplifiers (say, in a quad package) to derive reference voltages and switch them to V_{cc} to provide simple range changing.

Another convenient technique for low level signals is that shown in Fig. 4.2(f). Amplifier A_1 is used as a fixed-gain preamplifier, and the input offset is nulled by application to A_2. The advantage of this scheme is that low-level inputs, including A_1 intrinsic offset voltage, can be balanced at A_2 with voltage levels multiplied by the pre-amplifier gain, since input referred voltage is V_o/G_1. In addition, optimised resistance balance of A_1 is easily implemented.

A word of warning here. When designing even simple circuits,

CONSIDER TOLERANCES

A typical example is shown in Fig. 4.3. The nominal range of adjustment (V_2 to V_1) is ± 0.48 V. The value of V_1 calculated with the tolerances shown is

$$V_1 = -10 + 20 \times \frac{11\text{K}4}{11\text{K}4 + 9\text{K}8} = +0.75$$

and $\qquad V_2 = -0.38.$

The resulting range of control is therefore very much different from that desired.

An alternative to (d) for non-inverting applications is shown in Fig. 4.2(g). The point to remember here is that R_4 should be much smaller than R_1 (e.g. if $R_1 = 10$ kΩ, $R_4 = 100 \,\Omega$), otherwise the feedback

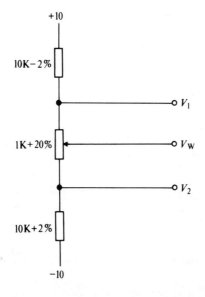

Fig. 4.3 The effects of resistor tolerances

fraction will be changed by adjusting the potentiometer, and zero and gain controls will be interactive. If $R_3 \gg R_4 \ll R_1$, gain to $V_{in} = 1 + R_2/(R_1 + R_4)$, and input voltage offset generated $\approx$

$$\pm \frac{V_w R_4}{R_4 + R_3} \frac{R_2}{R_1 + R_2}.$$

The voltage follower circuit shown in Fig. 4.2(h) is a simplification of (g), and for near-unity gain, $R_1 \gg R_2$ (i.e. $\beta \approx 1$). Range of control would be $\approx \pm V_w R_2/R_1$. The circuit in Fig. 4.2.(i) is often helpful when attempting to balance the amplifier input terminal resistance to maximise common-mode rejection.

4.2.2 Amplifier balance terminals

The terminals provided on many amplifiers are very useful for eliminating intrinsic offset errors, but the range of adjustment is usually limited to about $\pm 15\,\text{mV}$. The specification for offset error is given as $\pm x\,\text{mV}$, so it is to be expected that there will be a production spread in any family of devices, and some amplifiers may well have virtually zero error. The reason for the existence of the input error is obvious from Fig. 4.4. Both bipolar inputs (a) and FET inputs (b) are subject to differences between T_1 and T_2. If the two input terminals are at the same voltage, then $V_{be_1} = V_{be_2}$. Ideally, V_3 should equal V_4, but because of small differences between gm_1 and gm_2, the currents into the loads are not identical.

A rigorous fundamental analysis of the differential amplifier has been given by Middlebrook.[13] Any differential arrangement, and recent amplifier designs include complex arrangements of npn, pnp,

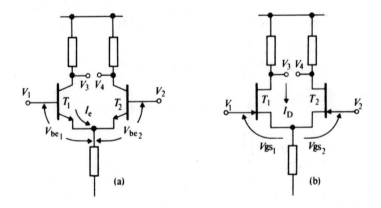

Fig. 4.4 The source of amplifier input offset voltage

'lateral transistors', 'super β transistors', FETs and Darlington pairs, etc., is likely to have an inherent mismatch betwen devices. Basically, the match obtainable with integrated circuits is far superior to that obtained by selecting discrete transistors, largely due to the simultaneous diffusion of each integrated device, and the close proximity of devices on a small-area chip minimising thermal gradients. Some idea of the quality of the match can be demonstrated by considering that typical $V_{be_1} - V_{be_2}$ values of 1 mV for identical device currents are common, whereas the actual magnitudes of V_{be} may be about 600 mV.

Considering the bipolar input stage, for any device

$$I_E = I_S(e^{\frac{q V_{be}}{KT}} - 1) \qquad \ldots [63]$$

where I_S = thermally generated junction leakage current
q = electronic charge
K = Boltzman's constant
T = Absolute temperature

$$\therefore V_{be} \approx \frac{KT}{q} \log_e \frac{I_E}{I_S} \qquad \ldots [64]$$

Since K, T and q are physical constants, and I_E is a defined value, it is I_S, the leakage current, which determines the variations of V_{be} between devices operating at identical currents; I_S is a function of the device construction.

Temperature dependence of V_{be} is thus

$$\frac{dV_{be}}{dT} = \frac{V_{be}}{T} - \frac{KT}{qI_S}\frac{dI_S}{dT} \qquad \ldots [65]$$

Equation [65] is shown[12] to be $\approx -2.2\,\text{mV/}^\circ\text{C}$. Recent integrated circuits have offset voltage drift capabilities down to 1 μV/$^\circ$C,* so the matching or internal temperature compensation is capable of reducing drift more than three orders of magnitude. Also,

$$\frac{d V_{offset}}{dT} = \frac{d V_{be_1}}{dT} - \frac{d V_{be_2}}{dT} = \frac{V_{be_1} - V_{be_2}}{T} \qquad \ldots [66]$$

so it is clear that matching V_{be} values reduces drift.

Now it might be assumed from the foregoing that using the amplifier offset control terminals to balance the effects of input offset error would reduce the offset drift. This is not necessarily the case, because it depends upon the method by which the offset error is balanced; in many cases, nulling the amplifier by the terminals

---

* Chopper, commutated auto zero and multi-element transistors can improve on this performance.

provided can make the drift worse. Always check carefully to make sure that using the balance terminals does not degrade the drift when this latter parameter is important. Table 4.3 shows the likely effects of utilising balance terminals on some common devices. For those devices where drift is degraded by offset adjustment, it is typically increased by about $3\,\mu V/\,^\circ C$ per 1 mV of offset adjustment.

However, the larger the basic device drift, the more it will swamp the increased drift effect caused by the nulling process; so that if we start with an amplifier which has a typical drift of $10\,\mu V/\,^\circ C$ when the offset voltage is initially zero, nulling an amplifier with an initial 1 mV error may degrade the drift to about $13\,\mu V/\,^\circ C$. However, with an amplifier such as the 725, the offset drift could be $1\,\mu V/\,^\circ C$ with initial offset of zero, or $10\,\mu V/\,^\circ C$ with an initial offset of 3 mV; nulling the 3 mV amplifier would not necessarily produce $1\,\mu v/\,^\circ C$ shift but there could be a residual drift introduced by the nulling process of about $1.5\,\mu V/\,^\circ C$, resulting in a final drift value of $2.5\,\mu V/\,^\circ C$ (which, nevertheless, is lower than the initial unbalanced value).

Another important point to consider with regard to offset nulling, is that the open-loop gain of many devices changes with the offset applied. In high-accuracy applications this could be an important fact; the open loop gain of a 741-type op amp has been found to vary by a factor of more than 10 for offset adjustments of ± 7.5 mV.

Considering the FET input stage of Fig. 4.4(b), for typical devices,

$$I_D = I_{DSS}\left(1 - \frac{V_{GS}}{V_p}\right)^2 \qquad\qquad \ldots [67]$$

Table 4.3 *Balance effects on drift.*

Drift could be made worse	Drift probably reduced	Little effect on drift
μA741	μA725	LF156
LF152	LM121	LM118
LM112	OP-05	AD504
μA709	AD510	AD508
LF155		
UC4250	SSS725	OP-15
BB3510	LM194	BB3527
OP-02	LM10	μA714
CA3140		OP-20
BB3500		
AD517		

I_{DSS} = drain current for $V_{GS} = 0$

V_p = pinch-off voltage, i.e. V_{GS} value for $I_D = 0$

$$\therefore V_{GS} = V_p \left(1 - \sqrt{\frac{I_D}{I_{DSS}}} \right)$$

The matching of V_{GS} at a specified drain current can be defined by measuring V_p and I_{DSS} for different devices; however, whereas a match of 1 mV is relatively easy for bipolar devices, 10 mV is more difficult to achieve with FETs. Also another difficulty arises with MOSFET devices used for linear inputs (e.g. CA3130), in that they have time-dependent V_{GS} variations under continuously biased conditions. It can be shown that

$$\frac{dV_{GS}}{dT} = -2.2 \, \text{mV/}^\circ\text{C} + \frac{7 \times 10^{-3} I_D}{g_{fs}} \qquad \dots [68]$$

where g_{fs} = forward transconductance. It is therefore theoretically possible to achieve zero change of V_{GS} with temperature by selection of I_D and g_{fs}; in practice there are other variable parameters which become effective at low levels, thus preventing the matching which would seem to be possible, and matched-pair offset drifts of $10 \, \mu\text{V/}^\circ\text{C}$ are rarely bettered. In some devices internal current-balance compensation is utilised to improve the drift values.

Each millivolt of offset error causes about $3.5 \, \mu\text{V/}^\circ\text{C}$ offset drift error, and in many cases external nulling (using terminals provided) degrades the drift; however, some recent developments (e.g. LF156) include specially designed balance circuits such that external balance adjustments have little effect on drift.

In every case where potentiometers are used, low temperature coefficients of resistance should be selected, as poor potentiometers can further degrade the drift. For best performance, the measured resistance values existing each side of the wiper after balancing should be replaced by two low temperature coefficient fixed resistors.

4.3 Variable gain controls

For the simple amplifier there are many options. Figure 4.5 shows some of the possibilities. Those shown at (a), (e), and (h) are linear, but the others are non-linear with potentiometer position.

One of the most important points to consider is the effect of the gain control on the input offset error. Since the input bias current flows into each input terminal, changing the terminal resistance will produce a change in the current generated input offset voltage. Consider the simple example in Fig. 4.6(a).

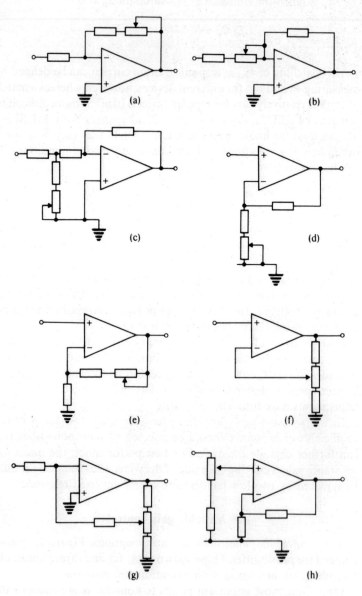

Fig. 4.5 Gain adjustment circuits

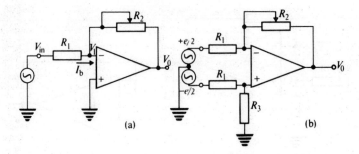

Fig. 4.6 Offset change with gain (a) and differential input (b)

The input bias current I_b flows from a Thevenin equivalent source resistance of $R_1 R_2/(R_1 + R_2)$. This must generate an input voltage of $I_b R_1 R_2/(R_1 + R_2)$ at V_1; hence if R_2 is changed to effect a gain adjustment, V_1 must change and any initially nulled offset error will change. Of course, the magnitude of the change may not be relevant in some circuits, especially when FET input devices are used. It is important not to forget, however, that FET bias (leakage) current approximately doubles for every 10 °C increase of temperature, whereas input bias currents for bipolar amplifiers tend to decrease with temperature. The great advantage of using a circuit such as Fig. 4.5(g) is that the feedback resistor effectively isolates resistance change from the inverting input terminal. The potential divider can be of relatively low resistance (compatible with output loading limitations) and the feedback resistor much higher in value, thus minimising the effects of gain and zero interaction. The other advantage of this circuit is that high closed-loop gains can be achieved without recourse to very high resistance values.

The gain of the differential input can be derived by considering the split source shown in Fig. 4.6(b).

$$V_o = -\left(+\frac{e}{2}\frac{R_2}{R_1} \right) + \left(-\frac{e}{2}\frac{R_3}{R_1 + R_3}\frac{1}{\beta} \right)$$

where
$$\beta = \frac{R_1}{R_1 + R_2}$$

$$\therefore V_o = \frac{-R_2}{R_1}\frac{e}{2} - \frac{R_3(R_1 + R_2)}{R_1(R_1 + R_3)}\frac{e}{2} \qquad \ldots [69]$$

If $R_3 = R_2$ then $\quad V_o = \frac{-R_2}{R_1}e$ as expected.

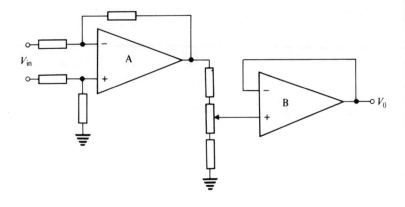

Fig. 4.7 Alternative gain adjustment technique

From [69],   the gain is $- V_o/e =$

$$\frac{R_2}{2R_1}\left(1 + \frac{R_3}{R_1 + R_3}\right) + \frac{R_3}{2(R_1 + R_3)} \qquad \ldots [70]$$

so it can be seen that the differential gain expression is no longer dependent directly on R_2 as with Fig. 4.6(a). A much more satisfactory arrangement would be that in Fig. 4.5(g).

The difficulty of adjusting d.c. gain whilst at the same time minimising zero offset shift can often be overcome by using an arrangement like that of Fig. 4.7. Amplifier A represents a fixed-gain preamplifier where drift and common-mode rejection can be optimised, and the output is connected to an attenuator. The attenuator can then drive B which may either be a follower as shown or can have extra gain as required. Any errors produced by B are then referred to the input terminals divided by gain A, and the attenuator can be a low-resistance chain, thus generating minimum offset to amplifier B with adjustment. The main consideration with this circuit is the dynamic range of the source signal and whether or not amplifier A can cover the full range required without saturation.

Another alternative for Fig. 4.6(b) is to use ganged potentiometers with one half in series with the feedback loop and the other from the non-inverting input to ground; this is often not very convenient.

The gain of the well-known 'instrumentation amplifier', Fig. 4.8, can be adjusted by the use of the single resistor R_3 without having a significant effect on offset if $R_3 \ll R_1$. The gain can again be calculated by looking at each half of the input separately as with the conven-

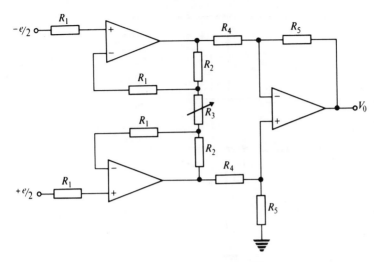

Fig. 4.8 Instrumentation amplifier

tional differential input. The overall gain then becomes

$$V_o = \frac{R_5}{R_4} \left(\frac{e}{2} \left[\frac{R_2 + R_3/2}{R_3/2} \right] + \frac{e}{2} \left[\frac{R_2 + R_3/2}{R_3/2} \right] \right)$$

$$\frac{V_o}{e} = G = \frac{R_5}{R_4} \left(1 + \frac{2R_2}{R_3} \right) \qquad \dots [71]$$

However, the gain is a non-linear function of R_3 as in Fig. 4.5(d), and demonstrated by [71]. The most practical solution is to make R_3 a fixed value and then vary the gain by following the preamplifier by a stage such as Fig. 4.9.

For a further discussion of other gain determining configurations, see reference 14. For each of the above expressions for gain, differentiating with respect to the relevant variable(s) will show how gain can vary with the temperature dependence of the resistors. For example, from [71],

$$dG = \frac{\partial G}{\partial R_5} dR_5 + \frac{\partial G}{\partial R_4} dR_4 + \frac{\partial G}{\partial R_2} dR_2 + \frac{\partial G}{\partial R_3} dR_3$$

$$\frac{dG}{G} = \frac{dR_5}{R_5} - \frac{dR_4}{R_4} + \frac{2R_2}{2R_2 + R_3} \frac{dR_2}{R_2} - \frac{2R_2}{2R_2 + R_3} \frac{dR_3}{R_3} \qquad \dots [72]$$

Writing percentage changes as G' and R', then percentage change of gain

$$G' = R_5{}' - R_4{}' + \frac{2R_2}{2R_2 + R_3} (R_2{}' - R_3{}')$$

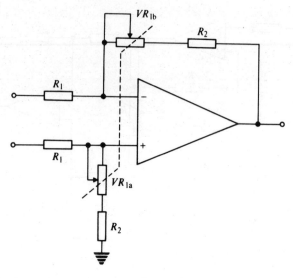

Fig. 4.9 Ganged potentiometer adjusts differential amplifier gain and minimises offset change

Many transducers, especially strain-gauge types, are connected to form a simple bridge. With a four-active arm configuration, this maximises output at the same time as providing compensation for temperature coefficient of resistance and apparent strain caused by differential linear expansion of gauge to substrate. In these applications a simple gain control could be provided by using a tracking voltage regulator to supply the bridge, as in Fig. 4.10. In this way a single resistor controls the $+V$ and $-V$ in a tracking mode, so that $+V = -V$ in magnitude. The output of a four-arm active bridge is given by $\dfrac{\Delta R}{R} 2 V$, where $\dfrac{\Delta R}{R}$ is the fractional change of resistance for each arm. Then, if $R_1 \gg R$,

$$V_o = \frac{R_2}{R_1 + R_s} \frac{\Delta R_2}{R} 2V \qquad \ldots [73]$$

where $R_s = \dfrac{\text{bridge output resistance}}{2}$

Thus V_o is proportional to V, and changes of sensitivity can be implemented by adjusting the single resistor VR. With low-resistance bridges this requires the regulator to provide adequate regulation under the changing current load conditions. The great advantage of using balanced supplies is that the common-mode output from the bridge is approximately $0\,V$ and remains at $0\,V$ for sensitivity

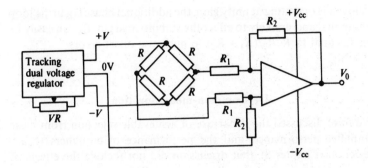

Fig. 4.10 Sensitivity adjustment of a bridge transducer

adjustments, thus optimising common-mode rejection performance. Another alternative, that of inserting a resistor in series with the bridge and adjusting the bridge supply by dropper from a constant-voltage source, is often not to be recommended due to bridge resistance temperature-dependence affecting the bridge supply voltage (unless temperature compensation is incorporated or, for example, when strain gauges are used, the gauge factor temperature-dependence is equal and opposite in sign to the bridge resistance temperature-coefficient).

Another very convenient method for gain adjustment is that shown in Fig. 4.11. Resistors R_1 and R_2 can be selected for a given gain range required, and then R_3, R_4 and VR_1 chosen to give the same percentage variation within each range as required. There is no loading effect on the potentiometer due to the presence of the voltage follower, and the ratio $R_1/(R_1 + R_2)$ is accurately defined. Further, since the voltage

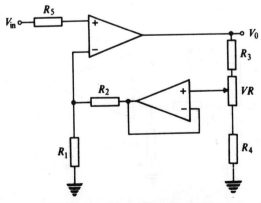

Fig. 4.11 Buffered gain adjustment

follower is operating at unity gain, the additional phase lag in the loop is minimal and unlikely to effect the stability margin. R_5 can easily be made equal to $R_1 R_2/(R_1 + R_2)$ to minimise offset errors.

4.4 Common-mode rejection

We have discussed the departure of achievable rejection from basic amplifier performance, and the performance of amplifiers to a.c. rejection (Chapter 3); that discussion did not include the effects of resistor tolerance errors on d.c. rejection. Refer to Fig. 4.12. a is the fractional error in resistance value; for example, if the error is $+1$ per cent, then $a = +0.01$. Observation indicates that with the errors allocated as shown, the amplifier differential input is maximised. Then

$$V_1 = \frac{R_2(1+a)}{R_2(1+a) + R_1(1-a)} V_{in}$$

$$= \frac{R_2}{R_2 + R_1 (1-a)/(1+a)} \qquad \dots [74]$$

$$V_2 = \frac{R_2(1-a)}{R_2(1-a) + R_1 (1+a)} V_{in}$$

$$= \frac{R_2}{R_2 + R_1 (1+a)/(1-a)} \qquad \dots [75]$$

The differential gain to $V_1 - V_2 \approx \dfrac{1}{\beta}$

$$= \frac{R_1(1-a) + R_2(1+a)}{R_1(1-a)} = \frac{R_1 + R_2(1+a)/(1-a)}{R_1} \qquad \dots [76]$$

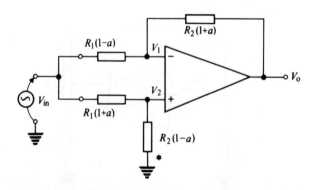

Fig. 4.12 Resistor tolerance effects on common-mode rejection

From [74], [75] and [76],

$\dfrac{V_o}{V_{in}} = G_{cm}$, common-mode gain,

$$= \frac{R_1 + R_2(1+a)/(1-a)}{R_1}\left[\frac{R_2}{R_2 + R_1(1-a)/(1+a)}\right.$$

$$\left. - \frac{R_2}{R_2 + R_1(1+a)/(1-a)}\right]$$

$$= \left(1 + \frac{R_2(1+a)}{R_1(1-a)}\right)\left[\frac{R_1 R_2\dfrac{1+a}{1-a} - \dfrac{1-a}{1+a}}{\left(R_2 + R_1\dfrac{(1-a)}{(1+a)}\right)\left(R_2 + R_1\dfrac{(1+a)}{(1-a)}\right)}\right]$$

If a is small,

$$G_{cm} = \left(1 + \frac{R_2}{R_1}\right)\left[\frac{4aR_1R_2}{(R_1+R_2)^2}\right] \qquad \ldots [77]$$

So for any given fractional error, the maximum common-mode gain can be calculated (i.e. for a random combination of fixed-tolerance resistors, the worst-case common-mode error output). Further, if the differential gain is high, i.e. $R_2 \gg R_1$, then [77] becomes

$$G_{cm} \approx 4a \qquad \ldots [78]$$

This means that for a 1 V common-mode input signal, with $a = 0.01$ (i.e. all resistors on their 1 per cent tolerance limits), the resulting error output is 40 mV. We can therefore define the common-mode rejection ratio due to resistor errors alone as

$$\text{CMRR} = \frac{\text{differential gain}}{G_{cm}} = \frac{R_2/R_1}{G_{cm}} = \frac{R_2/R_1}{4a} = \frac{R_2}{4aR_1}$$
$$\ldots [79]$$

Note that the above expression contains nothing to describe the amplifier generated error. By definition, the amplifier common-mode rejection ratio is $\text{CMRR}_A = \dfrac{R_2/R_1}{G_{cmA}}$, so we have two separate ratios generating error voltages. The amplifier input terminals 'see' a common-mode input voltage of very nearly

$$\frac{R_2}{R_1 + R_2}V_{in} \text{ as } V_1 \approx V_2.$$

Again, if R_2/R_1 is large, then V_{cm} (input) $\approx V_{in}$.

The total output error in the worst case is therefore

$$V_o = G_{cm} V_{in} + G_{cmA} V_{in}$$

$$\therefore V_o = \frac{R_2/R_1}{CMRR} V_{in} + \frac{R_2/R_1}{CMRR_A} V_{in}$$

Overall common-mode gain of the resistor/amplifier combination,

$$G_{cmo} = \frac{V_o}{V_{in}} = \frac{R_2/R_1}{CMRR} + \frac{R_2/R_1}{CMRR_A}$$

and the overall common-mode rejection ratio is

$$CMRR_o = \frac{R_2/R_1}{G_{cmo}} = \frac{CMRR \times CMRR_A}{CMRR + CMRR_A} \qquad \cdots [80]$$

The individual ratios therefore add as though they were resistors in parallel.

Let us consider an example. Let closed-loop gain $R_2/R_1 = 100$. For 1 per cent resistors, the worst case gives $CMRR = 100/4a = 2.5 \times 10^3$. Let the amplifier have a common-mode rejection ratio of $100 \, dB = 10^5$. Then from [80], the effective $CMRR_o = \frac{1}{4.1} \, 10^4$ $\approx 68 \, dB$. The unfortunate result of using unmatched 1 per cent resistors could reduce the specified rejection ratio of an amplifier from 100 dB to 68 dB!

In order to exploit the high performance capability of op amps, it is necessary either to use well-matched resistors, or to apply a common-mode input signal and trim one resistor value to minimise V_o. This is usually done by including a potentiometer in series with the resistor * in Fig. 4.12, so that the initial resistor value is lower than the optimised value, and then adjusting for best performance. However, be warned: if each resistor is, say, a metal-oxide type, it could well have a temperature coefficient of up to $\pm 200 \, ppm/^\circ C$. This means that over a 100 °C temperature range, each resistor could change its value by 2 per cent, so that rejection ratio may degrade from its optimum setting to something very much worse at temperature extremes. So here we require resistors either with tracking temperature coefficients, or with very low absolute coefficients.

For readers interested in transducer bridge applications, the bridge outputs are calculated in Appendix 5. There is another point to consider when designing bridge circuits with the objective of maximising CMRR. In Fig. 4.13 it is inferred that the bridge common-mode output is 0 V if $V+ = V-$. This is often true for well-matched transducer bridge arms, but is not necessarily always true.

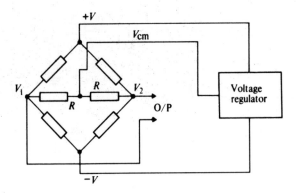

Fig. 4.13 Bridge common-mode restoration

The most important point, however, is that V_{cm} may not be constant with temperature, especially where series-dropping resistors and/or temperature-compensation thermistors are utilised. Hence errors will arise as a result of changing V_{cm} with temperature of magnitude dependent on the rejection capabilities of the rest of the circuit. Figure 4.13 shows a method by which these errors could be minimised. The common-mode bridge output V_{cm} is monitored, and R chosen so as not to load the bridge output. V_{cm} will not change if the bridge output is differential. The voltage regulator is then controlled in such a way that $(+V)+(-V)$ is not equal to zero but is the required value to restore V_{cm} to zero. For example, initially $+V=+10\,V$ and $-V=-10\,V$, so that $V_{cm}=0$ and the bridge total excitation is 20 V. Over temperature, V_{cm} would shift to $+1\,V$ with constant supplies, therefore $(+V)$ should become $+9\,V$ and $(-V)$ should become $(-11\,V)$ to maintain $V_{cm}=0$ with the total bridge supply $(+9)-(-11)$ constant at 20 V.

A final comment on rejection ratios. They are normally expressed in voltage dB, and it may sometimes be desirable to add ratios directly in dB. This of course represents a multiplication of the original ratios due to the logarithmic derivation of dB.

Any gain-adjustment circuit which operates in the presence of an impressed common-mode voltage is likely to cause difficulty, so care is required in the implementation. A circuit such as Fig. 4.14(a) is a possible method by which common-mode rejection can be maximised for driving floating loads; adjustment of R_4 for gain control will not have a significant effect on resistive imbalance at the input terminals, and common-mode voltages will be equal in magnitude at each end of R_4, so there will be no current flow through R_4 due to that source. The gain of such a configuration could be obtained as below; consider

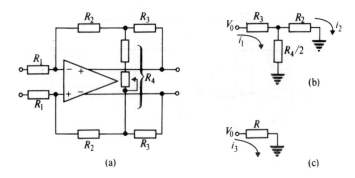

Fig. 4.14 Gain adjustment of differential-output amplifier

each half of the amplifier separately, then the feedback path to V_o is as shown in Fig. 4.14(b), the amplifier input terminal becoming a virtual earth point. All we need to do is calculate the equivalent resistor R Fig. 4.14(c), then the gain is simply R/R_1.

$$i_1 = \frac{V_o}{R_3 + \dfrac{R_2 R_4/2}{R_2 + R_4/2}} \qquad i_3 = \frac{V_o}{R}$$

$$i_2 = \frac{R_4/2}{R_2 + R_4/2} i_1 = \frac{V_o R_4/2}{R_3(R_2 + R_4/2) + R_2 R_4/2}$$

For the resistance values to equate in (b) and (c), $i_2 = i_3$.

$$R = \frac{R_3(R_2 + R_4/2) + R_2 R_4/2}{R_4/2} = R_2 + R_3 + \frac{2R_2 R_3}{R_4}$$

$$\therefore \text{ Gain } G = \frac{R_2}{R_1} + \frac{R_3}{R_1} + \frac{2R_2 R_3}{R_1 R_4} \qquad \dots [81]$$

4.5 Single supply circuits

It is often desirable to design amplifiers for use on a single power supply line, e.g. for battery operation. Clearly a simple possibility is to derive a common reference half-way between 0 and $+V_{cc}$, and there are many ways of doing this. Some typical examples are shown in Fig. 4.15. The alternatives to using a derived common line are either to use special 'ground sensing' amplifiers described in Chapter 9, or to reference the load to a derived voltage level somewhere between the 0 and $+V_{cc}$ lines. The simplest example of the latter is shown in Fig. 4.16, which in its basic form would be identical to the use of a common rail. The reference circuit could be any of the examples from

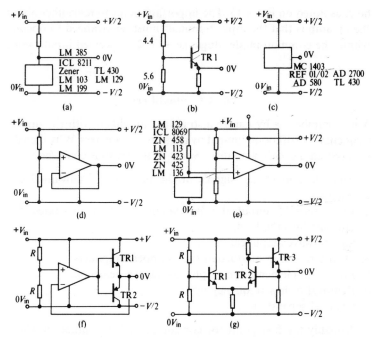

Fig. 4.15 Derivation of balanced outputs from a single supply

Fig. 4.15. The criteria are low source impedance and low temperature dependence so that the errors generated by the load current and temperature changes are negligible compared with the required output. A convenient method for minimising the effects of bridge common-mode changes as discussed in section 4.4 can be implemented by sensing the input common-mode voltage and using it to control the reference voltage output. There are a number of integrated-circuit reference devices on the market which can be used

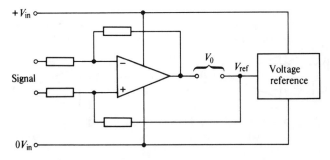

Fig. 4.16 Amplifier operation from a single supply

here, as noted on Fig. 4.15. The important point to remember about the op amp is that the input terminals must be arranged to operate within the common-mode range; i.e. $+ V/2 \pm$ specified input range.

4.6 A.C. amplifiers

A d.c. amplifier is by nature also an a.c. amplifier, with the upper-frequency $- 3\,dB$ point limited by the same parameters. Why then do we need an a.c. *only* amplifier? Some of the reasons could be:

1. Amplification of a.c. signals in the presence of unwanted d.c. voltages.
2. Prevention of amplified offset voltages driving unwanted d.c. currents into the load.
3. Square-wave amplification.
4. Drift removal in modulated (i.e. chopper) amplifiers.
5. Ease of implementation of phase-sensitive detection.
6. Removal of low-frequency noise.
7. Channel separation.

Probably the first point for consideration is the situation where high a.c. gain is required. There is a difference in principle between: (a) an a.c. coupled amplifier, where the coupling appears either at the input, or output, or both, and the amplifier itself has equal a.c. and d.c. gain; and (b) the situation where a.c. coupling may or may not be present but the amplifier itself has a defined low-value d.c. gain and a high-value a.c. gain. The alternatives can be seen in Fig. 4.17(a) and (b). In (a) we have the situation where the signal source and the load are a.c. coupled, but the a.c. and d.c. gains within the box are the same. In this case, high d.c. source voltages are removed, and amplifier drift and offset voltage is removed from the output. The amplifier d.c. and

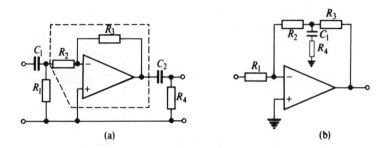

(a) (b)

Fig. 4.17 Simple a.c. amplifiers

a.c. gain is

$$\frac{R_1 R_3}{R_1 R_2 + R_s(R_1 + R_2)}$$

where R_s is the source resistance. Resistor R_1 may not always be necessary; it has been shown here to indicate that the leakage current of C_1 must be considered, as R_4 is necessary to prevent accummulation of charge on C_2. Bandwidth limitation by C_1 would be defined from

$$G = \frac{j\omega C_1 R_1 R_3}{(R_1 + R_2) + j\omega C_1 R_1 R_2} \text{ if } R_s \text{ is negligible}$$

or, if $R_1 \gg R_2$, which can usually be arranged,

$$G = \frac{R_3}{R_2 + 1/j\omega C_1} = \frac{j\omega C_1 R_3}{1 + j\omega C_1 R_2}$$

$$|G| = \frac{\omega C_1 R_3}{\sqrt{1 + \omega^2 C_1^2 R_2^2}} = \frac{R_3}{\sqrt{2} R_2} \text{ at } -3\,\text{dB}$$

$$\therefore \omega_o = \frac{1}{R_2 C_1} \qquad \dots [82]$$

The bandwidth limitation by C_2 is of course $\omega_o = 1/C_2 R_4$. In practice C_1 and C_2 can be made sufficiently large so that the low-frequency response is as low as required.

Figure 4.17(b) is an example of a circuit with defined high a.c. gain but low d.c. gain. This is a convenient circuit where d.c. coupling is desired without the effects of a high d.c. offset voltage. The d.c. gain is $(R_2 + R_3)/R_1$, and the high-frequency a.c. gain is

$$G = \frac{R_2}{R_1} + \frac{R_3}{R_1} + \frac{R_2 R_3}{R_1 R_4},$$

using the equivalent-resistance method calculated for [81]. The same method can be used to obtain the lower $-3\,\text{dB}$ point.

$$G = \frac{R_2 + R_3}{R_1} + \frac{R_2 R_3}{R_1 (R_4 + 1/j\omega C_1)}$$

$$G = \frac{R_2 + R_3}{R_1} + \frac{\omega^2 C_1^2 R_2 R_3 R_4}{R_1 (1 + \omega^2 C_1^2 R_4^2)} + \frac{j\omega C_1 R_2 R_3}{R_1 (1 + \omega^2 C_1^2 R_4^2)} \qquad \dots [83]$$

from which the magnitude of G may be calculated and the cutoff frequency obtained.

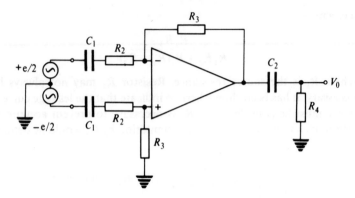

Fig. 4.18 Differential-input a.c. amplifier

A differential-input, a.c. coupled amplifier may look like that shown in Fig. 4.18. Assume the effect of C_2 is small.

The output is therefore

$$V_o = -\frac{R_3}{R_2 + 1/j\omega C_1}\frac{e}{2} - \frac{R_2 + R_3 + 1/j\omega C_1}{R_2 + 1/j\omega C_1}\frac{R_3}{R_3 + R_2 + 1/j\omega C_1}\frac{e}{2}$$

$$-V_o = e\frac{R_3}{R_2 + 1/j\omega C_1}$$

and the bandwidth is identical to [82].

A combination of circuits Fig. 4.17(a) and (b) may be used in some circumstances. Figure 4.19 shows a possible arrangement for a non-inverting application. Assuming C_1 and C_3 do not cause bandwidth limitation, the low-frequency gain is given by

$$G = \frac{R_2 + R_3}{R_2} = 1 + \frac{R_3}{R_2}$$

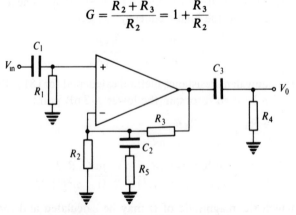

Fig. 4.19 Non-inverting a.c. amplifier

In many applications, R_2 may be eliminated so that the low-frequency gain is unity. The a.c. gain is given by

$$G = \frac{R_3[1 + j\omega C_2(R_2 + R_5)] + R_2(1 + j\omega C_2 R_5)}{R_2(1 + j\omega C_2 R_5)}$$

$$= 1 + \frac{R_3}{R_2} \frac{1 + j\omega C_2(R_2 + R_5)}{(1 + j\omega C_2 R_5)} \qquad \ldots [84]$$

At $-3\,\text{dB}$, $\qquad |G| = \frac{1}{\sqrt{2}}\left(1 + \frac{R_3}{R_2}\right) \qquad \ldots [85]$

Hence by equating [84] and [85] the $-3\,\text{dB}$ point can be obtained.

4.6.1 D.C. restoration

It is often desirable to amplify a square or pulsed waveform and then to integrate or filter the output to provide a d.c. level, for example, in synchronous detection. If half-wave synchronous detection is used, half the usable signal is lost simply by the lack of d.c. restoration through a.c. couplings. The d.c. level can be easily restored by using an 'active diode' circuit to clamp the output to the desired polarity. Figure 4.20 shows a typical arrangement. During the negative output overshoot, the amplifier A_1 output terminal goes positive as a result of the negative signal being applied to the inverting input by R_f, and D_2 turns on hard. As a result of D_2 being inside the feedback loop, the effective output is clamped very near to 0 V, and the temperature coefficient of the diode drop effect is reduced by $1/(1 + \beta A)$. For

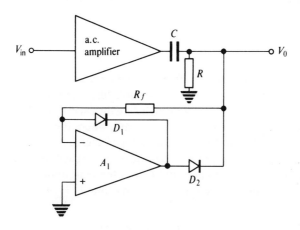

Fig. 4.20 D.C. restoration

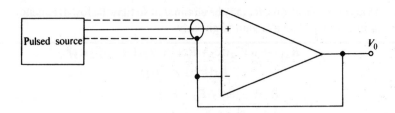

Fig. 4.21 Screen driving

positive-going outputs the amplifier A_1 output goes negative, D_2 turns off and D_1 turns on, clamping the output of A_1 so that it does not saturate. The loading on C is thus R and R_f in parallel. In most applications resistor R is not required as any leakage current from C will flow through R_f.

4.6.2 Screen driving

Some instrument applications require the transmission of a pulsed waveform along a screened lead; frequently the source is a high-resistance device, such that connecting the lead screen to ground produces unacceptable capacitive loading on the source. A voltage follower at the receiving end can be used to produce an active screen drive, as shown in Fig. 4.21. Since the screen is driven from a low-impedance source and is always at the same potential as the core, the core-to-screen capacity is effectively removed; protection against external electromagnetic radiation is still afforded, since induced currents in the screen flow harmlessly into the low driving source impedance.

4.6.3 Bootstrapping

This is a convenient point to familiarise the reader with the concept of 'bootstrapping', since it often causes difficulty on initial contact. The concept was originally introduced during the early days of valve time-base design for cathode ray tube displays, where there was a need to generate ramp waveforms with a high degree of linearity. The technique can be applied to resistors or capacitors, but it was originally in capacitor applications that the realisation was effected. In Fig. 4.22, the voltage across R at any time is $V_{in} + KV_o - V_o$ and therefore

$$i = \frac{V_{in} + V_o(K-1)}{R}$$

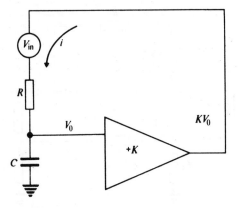

Fig. 4.22 Principle of 'bootstrapping'

If $K = +1$, then the current i is constant with value V_{in}/R. For the capacitor, $i = dq/dt = C\,dV_o/dt$, so that if i is constant $dV_o/dt = i/c$ and the slope of V_o is constant, producing a ramp of slope i/c. Now

$$i = \frac{V_{in} + V_o(K-1)}{R} = C\frac{dV_o}{dt}$$

$$\therefore \frac{dV_o}{dt} - \frac{(K-1)V_o}{CR} = \frac{V_{in}}{CR}$$

Solving for V_o gives

$$V_o = \frac{V_{in}}{1-K}\left(1 - e^{\frac{(k-1)t}{CR}}\right) \qquad \ldots [86]$$

If $K = 0$, i.e., the conventional charging of a capacitor through a resistor, the classic relationship results. The nearer K approaches unity, the larger the effective source voltage step of $V_{in}/(1-K)$. The effective time constant is $CR/(K-1)$, producing an effective capacitance (or resistance value) of $C/(K-1)$. Hence by making $K = +1$, we have an infinite time constant driven from an infinite voltage source and the output is a linear ramp. The deviation from ideal can be obtained from the ramp generation described by [86]:

$$\frac{dV_o}{dt} = -\frac{V_{in}}{1-K}\frac{K-1}{CR}e^{(K-1)t/CR}$$

$$= \frac{V_{in}}{CR}e^{(K-1)t/CR}$$

For large values of $K(\to 1)$ or small values of t, the initial slope is V_{in}/CR.

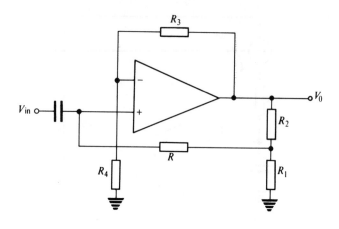

Fig. 4.23 Bootstrapped bias resistor R

Hence the percentage deviation from terminal linearity is given by

$$\frac{V_{in}/CR - (V_{in}/CR)e^{(K-1)t/CR}}{V_{in}/CR} \times 100$$

$$= 100(1 - e^{(K-1)t/CR}) \text{ percent} \qquad \dots [87]$$

The bootstrap principle can be applied to bias resistors on an amplifier input to increase the apparent value of the resistors and prevent their loading effect. One possible arrangement is shown in Fig. 4.23, where $R_2 \ll R \gg R_1$. Care must be taken to ensure that the overall negative feedback is greater than the positive feedback, or instability will result.

4.6.4 *VLF (very low-frequency) amplifiers*

There are many areas of application where long-period signals must be processed (e.g. $T = 50$ s), but direct coupled systems can cause difficulty as a result of transducer or amplifier drift. In such applications the low noise and stable performance of a bipolar preamplifier can be coupled with a long time-constant input a.c. amplifier. Figure 4.24 shows a possible arrangement. Amplifier G_1 is a bipolar low-noise, differential-input amplifier, and G_2 is connected as an a.c. amplifier. Transducer d.c. drift and G_1 drift is blocked by C_1, and the input a.c. time constant will be $C_1 \times$ amplifier input resistance in parallel with the effect of the bootstrapped R_1. If G_2 is a JFET or MOSFET input amplifier, the time constant can be made very large.

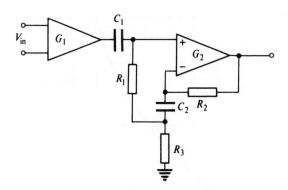

Fig. 4.24 Very low-frequency amplifier

For the a.c. coupling,

$$\frac{V_o}{V_{in}} = \frac{j\omega C_1 R}{1 + j\omega C_1 R} \qquad \left|\frac{V_o}{V_{in}}\right| = \frac{\omega C_1 R}{\sqrt{1 + \omega^2 C_1^2 R^2}}$$

where R is the effective input resistance.

At $-3\,dB$, $\qquad \left|\frac{V_o}{V_{in}}\right| = \frac{1}{\sqrt{2}} \qquad \therefore f_0 = \frac{1}{2\pi C_1 R}$

$$\therefore \left|\frac{V_o}{V_{in}}\right| = \frac{f/f_0}{\sqrt{1 + (f/f_0)^2}}$$

and the percentage error can be calculated. Chart 4.1 shows a plot of f_0/f against percentage transmission error,

$$\text{percentage error} = 100\left(1 - \frac{1}{\sqrt{1 + (f_0/f)^2}}\right) \qquad \ldots [88]$$

It can be seen that the $-3\,dB$ point of the a.c. coupling needs to be about 6.7 times lower than the frequency f of interest to reduce the transmission error to less than 1 percent. For an amplitude accuracy of 0.1 per cent, $f > 22.2 f_0$. If $f_0 = 1/2\pi C_1 R$ and the lowest frequency of interest is 0.02 Hz i.e. 50 s period), then the effective time constant CR of the coupling must be 177 s. If the value of C_1 is 1 μF (it will be imperative to use a low-leakage, non-electrolytic type), then the effective loading resistance must be greater than 177 MΩ, which represents a relatively easily obtainable value. Precautions must be taken with regard to circuit layout, protection against humidity etc. to maintain this level of performance.

With the advent of chip technology capable of producing bipolar

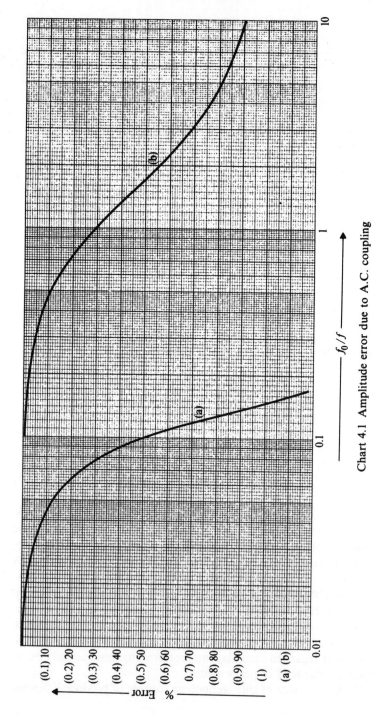

Chart 4.1 Amplitude error due to A.C. coupling

and MOS/JFET devices on the same circuit, very long input time constants are achievable, and the necessary low frequency response may be attainable with a single amplifier. However, there are still difficulties involved; it may not be possible to optimise noise, CMRR, capacitor leakage or other parameters, and the two amplifier approach offers greater flexibility. Hybrid devices (i.e. multichip) can be another useful method to achieve the overall result.

5

Noise and drift

5.1 General precautions

The subjects of noise and drift are extensive. Fundamentally, drift is really very low-frequency noise, so the whole forms one subject, the division between them being dependent on frequency content or bandwidth. However, there is one section which can be separated. Drift at constant temperature can be accurately described as noise, but drift of input voltages and bias currents with temperature can be considered separately. The theoretical aspects of these topics are adequately discussed in the literature, so the contribution here will be restricted to some of the practical aspects, and to an explanation of what it means to the designer. The typical frequency distribution of noise types is shown in Fig. 5.1.

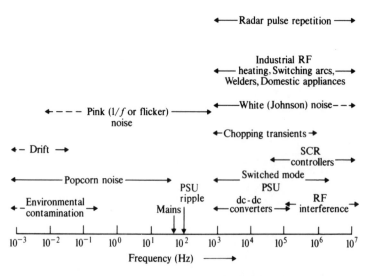

Fig. 5.1 Noise source frequency dependence

112

It should be remembered that an op amp does not need to provide gain at the frequency of radiated interference for the d.c. performance to be affected. The non-linear circuit devices are often capable of rectifying high-frequency pickup and changing the d.c. operating characteristics. Low-level circuits operated in industrial environments, for example, where RF heating or welding equipment is in use, will often be affected both by power-supply borne and radiated interference. In such cases, simple attempts at RF filtering will not prove successful unless all possible methods by which high frequencies can penetrate have been controlled.

Noise can therefore be either internally or externally generated. The internal noise mechanisms are largely controlled by the fundamental physics of the device, and the process of manufacture. Their values cannot be changed by the circuit designer, but their effects can be minimised by optimisation of circuit values and configuration. Externally generated noise can be minimised at source, and reduced by suitable enclosures and filtering. Added to these sources we have the power-supply induced noises: ripple, switching transients, line disturbance feedthrough, regulation, tracking errors, temperature dependence, etc.

Noise can be considered as being referred to the input terminals, irrespective of its source – this is convenient since all wanted signals can be referred to the input and signal-to-noise ratios established. It is present over a very wide frequency range as exemplified by Fig. 5.1, from d.c. to VHF. The bands tend to overlap, so that noise at any one frequency is a complex addition from many sources. Most noise generated externally tends to be repetitive, although equipment such as arc welders will produce repetitive bursts of random noise; the effects of such sources need to be reduced to insignificant levels before the full potential of an op amp can be realised. The magnitudes of input bias currents and offset voltages and their stabilities have improved enormously in the last few years, and so noise becomes an increasingly important source of error. It is important to be aware that op amps are usually tested on a statistical basis, so that published specifications will often not imply 100 per cent testing. Typically, there are guarantees that 90 per cent of all devices will meet the specifications, so it should not be assumed that selection will not be necessary to guarantee the performance of the completed circuit. The following points can be used as guidelines to circuit design:

1. Circuit bandwidth should be restricted to the minimum demanded by the signals to be processed.

Table 5.1 Some common noise sources

Source	Nature	Causes	Minimisation Methods
50 Hz Power (60 Hz)	Repetitive interference	Powerlines physically close to op amp inputs. Poor CMRR at 50 Hz. Power transformer primary-to-secondary capacitive coupling	Reorientation of power wiring. Shielded transformers. Single point grounding. Battery power. Twisted pairs. Avoid wiring loops
100 Hz Ripple (120 Hz)	Repetitive	Full-wave rectifier ripple on op amp's supply terminals. Inadequate ripple consideration. Poor PSRR at 100 Hz	Thorough design to minimize ripple. R–C decoupling at the op amp. Battery power
150 Hz (180 Hz)	Repetitive EMI	150 Hz radiated from saturated 50 Hz transformers	Physical reorientation of components. Shielding Battery power
Radio stations	Standard AM broadcast, through FM	Antenna action anywhere in system	Shielding. Output filtering. Limited circuit bandwidth
Relay and switch Arcing	High-frequency burst at switching rate	Proximity to amplifier inputs, power lines, compensation terminals, or nulling terminals	Filtering of HF components. Shielding. Avoidance of ground loops. Arc suppressors at switching source

Printed circuit board contamination	Random low frequency	Dirty boards or sockets	Thorough cleaning at time of soldering followed by a bakeout and humidity sealant
Radar transmitters	High frequency gated at radar pulse repetition rate	Radar transmitters from long-range surface search to short-range navigational – especially near airports	Shielding. Output filtering of frequencies $\gg$ PRR
Mechanical vibration	Random < 100 Hz	Loose connections, intermittent metallic contact in mobile equipment. Cable noise. Resistive, piezoelectric. Charged cables	Attention to connectors and cable conditions. Shock mounting in severe environments. Use of low-noise cables
Chopper frequency noise	Common-mode input current at chopping frequency	Abnormally high noise chopper amplifier in system	Balanced source resistors. Use bipolar input op amps instead. Use premium low-noise chopper

(*Courtesy of Precision Monolithics Inc.*)

2. Utilise the common-mode rejection capability of the amplifier wherever possible.
3. Use metal-film resistors in source and feedback networks.
4. Use select-on-test or high-tolerance fixed resistors in preference to pots.
5. Take care with physical layout, proximity of power lines and radiating sources.
6. Examine manufacturers' specifications carefully and make sure that differing devices are being compared on exactly the same bases.
7. Avoid ground loops.
8. Utilise input guarding to reduce capacitive and leakage noise pickup.
9. Solvent clean pc boards and then varnish or use some other suitable protection to reduce surface contamination.
10. Always keep in mind the environment in which the circuit has to operate, and whether or not special filtering is desirable.
11. Minimise source impedances.

A useful guide to device noise has been published by Precision Monolithics Inc.,[15] to whom I am grateful for permission to publish much of the information contained in Fig. 5.1 and Table 5.1.

Table 5.1 indicates some common sources of external noise

5.2 Pink noise (flicker noise, $1/f$)

The noise power is distributed logarithmically with frequency. This means that noise voltage or current, when referred to the input terminals, is equal in each decade of bandwidth. If the noise content in one decade is known, the total noise can be calculated. Since white noise contribution (see section 5.3) below about 10 Hz tends to be negligible, the decade 0.1 Hz to 1 Hz is often used as the basis for calculation. Noise voltage and current per square-root-cycle of bandwidth increase as the frequency decreases, so that the division between noise from this source and drift with time becomes difficult to determine, e.g., noise at 0.01 Hz (a period of 100 s) may well be considered as drift with time, but the effects of temperature must be defined. Figure 5.2 shows the general shape of a semiconductor device noise spectrum. The flicker noise below f_1 is usually attributed to leakage current and to contaminated or imperfect semiconductor surfaces. Normally, separate voltage and current noise levels referred to the input are specified to enable the designer to optimise or calculate signal-to-noise ratio based on any desired source resistance.

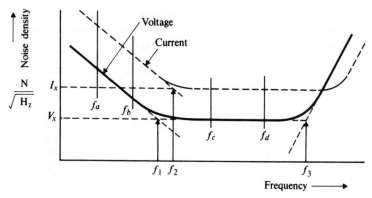

Fig. 5.2 Typical semiconductor device noise spectrum

Flicker noise voltage and current can be written as below:

$$V_{rms} = K_1 \sqrt{\ln\left(\frac{f_b}{f_a}\right)}$$

$$I_{rms} = K_2 \sqrt{\ln\left(\frac{f_b}{f_a}\right)} \qquad \text{where ln denotes } \log_e$$

where K is the constant derived from the 0.1 Hz to 1 Hz decade and f_b and f_a are the upper and lower frequencies of the band of interest. It can be seen that the lower f_1, the lower the total rms flicker noise will be for a typical small-signal amplifier. Hence f_1 can be considered as a figure of merit for low-noise amplifiers, with the best amplifiers having the lowest values. Table 5.2 gives some typical figures. Assuming that each noise component at each frequency is random, that is uncorrelated, the resulting amplitude distribution will be Gaussian in nature. This means that if the rms value is multiplied by 3, the resultant peak value will not be exceeded for approximately 99.73 per cent of the time. This leads us to another useful definition, that of 'crest factor', which is the ratio of peak to rms value, that is, 3 in this case.

The total voltage and current noise in the $1/f$ part of the spectrum between f_b and f_a is

$$V_{rms} = V_x \sqrt{f_1 \ln\left(\frac{f_b}{f_a}\right)} \qquad \dots [89]$$

$$I_{rms} = I_x \sqrt{f_2 \ln\left(\frac{f_b}{f_a}\right)} \qquad \dots [90]$$

Table 5.2 *Typical amplifier noise voltage*

Amplifier	Typical f_1 (Hz) $R_s = 0$	Voltage noise $nv/\sqrt{Hz} = x$	xf_1
μA714	4	10.5	42
SSS725	7	7	49
OP-07	6	10	60
739	15 (est.)	10	150
OP-02	10	20	200
AD504	20	9.5	190
SE5534	80	4	320
TDA1034	80	4	320
AD517	20	20	400
MC1741N	100 (typ.)	4.5	450
LM108	20–30	35	700
LF156	60	12	720
AD508	100	10	1 000
BB3510	100	12	1 200
LM148	20–30	60	1 200
LM725	300	7	2 100
741	200–500	20	4 000
LM10	100–200	45	4 500
TL064	500	40	20 000

with f_1 and f_2 having the values shown in Fig. 5.2 for the current and voltage noise density distribution, and V_x, I_x the noise density in $nV/\sqrt{Hz}$ and $nA/\sqrt{Hz}$ at f_1 and f_2 respectively.

For example, the computed total noise voltage for the LF156 over the frequency band 0.1 Hz to 30 Hz (Table 5.2) becomes

$$V_{rms} = 12 \times 10^{-9} \sqrt{60\ln300} = 0.22 \ \mu V \text{ rms}$$

If the noise is random, we can expect the peak-to-peak value to be less than 1.3 μV for more than 99 per cent of the time. Since the level of V_x can vary as well as f_1, an improved figure of merit could be given by $f_1 \times V_x$, as shown in Table 5.2.

Chart 5.1 shows the expected time proportionality of the relationship between peak and rms noise for a Gaussian-type distribution.

5.3 White noise

Noise power is distributed uniformly with frequency, i.e., each 1 Hz bandwidth has a constant noise power content. This is the type of noise which predominates in the band f_1 to f_3. The total white noise at the input of an amplifier is a combination of white-noise voltage, and

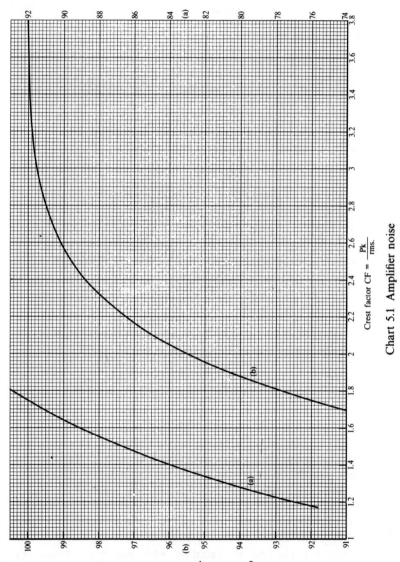

Chart 5.1 Amplifier noise

white-noise current which flows in the source resistance. The white-noise current is 'shot noise', which is attributable to the movement of electrons. White-noise voltage in the frequency band f_c to f_d is given by

$$V_W = V_x \sqrt{f_d - f_c} \qquad \ldots [91]$$

and white-noise current is

$$I_W = \sqrt{2q\, I_b(f_d - f_c)} \qquad \ldots [92]$$

where q = electron charge = 1.6×10^{-19} coulomb

I_b = bias current

f_d, f_c = upper and lower frequency limits.

As with voltage noise,

$$I_W = I_x \sqrt{f_d - f_c} \qquad \ldots [93]$$

The thermally generated noise in resistors is also white in nature, and is given by the well-known Nyquist expression,

$$V_T = \sqrt{4KTR\Delta f} \qquad \ldots [94]$$

where K = Boltzman's constant = 1.38×10^{-23} joules/K

T = absolute temperature

R = resistance value

Δf = bandwidth

It is interesting to note that although a differential input arrangement is desirable to maximise common-mode rejection, it usually makes the current-generated noise worse, since the balanced input arrangement implies the addition of voltage noise in both input terminals rather than one. Since the current-generated voltage noise is uncorrelated, the addition is the RSS value (RSS = root of the sum of the squares).

For a typical semiconductor, a high-frequency response is required for f_3 to be high with increasing source resistance; low-noise, high source resistance operation necessitates low collector currents, but current gain decreases at low currents, so that a compromise results. The increasing noise level beyond f_3 is rarely specified for voltage noise, since it tends to be at frequencies above which the amplifier has useful gain. However, sometimes the turn up is indicated for current noise (e.g. LM725). Some device data sheets present curves of total broadband noise and its variation with source resistance (e.g. TDA1034); this is useful information since it eliminates the need to carry out calculations for each of the noise components and then add them together.

The white noise generated in the source-connected resistors adds to

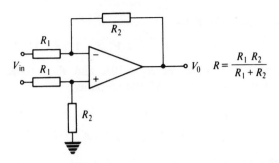

Fig. 5.3 Noise calculation resistors for a differential amplifier

the device noise, so that as source resistance increases, the total noise increases. Also the device noise current will produce increasing voltage noise at the input as the resistance increases. An article has been published by Smith and Sheingold,[16] which will be useful to readers wishing to pursue the subject more thoroughly. In Fig. 5.3 the source resistance in each terminal is $R_1 R_2/(R_1 + R_2) = R$. The current-generated white-noise voltage is therefore RI_w, and from [92],

$$V_R = R \sqrt{2q\, I_b (f_d - f_c)}$$

in each terminal. The total for both terminals is therefore

$$V_{R2} = R \sqrt{4q\, I_b\, (f_d - f_c)} \qquad \dots [95]$$

We can now add all of the noise components together to compute the resultant, over the total band f_a to f_3 in Fig 5.2. It must be remembered that the flicker and white-noise current generators produce voltages in each input terminal, as do the thermally generated resistor noises. The components then add on an RSS basis. We have, from [89],

$$V_{1\,\text{rms}} = V_x \sqrt{f_1 \ln\left(\frac{f_3}{f_a}\right)}$$

from [90], $$V_{2\,\text{rms}} = RI_x \sqrt{2f_2 \ln\left(\frac{f_3}{f_a}\right)}$$

from [91], $$V_{1w} = V_x \sqrt{f_3 - f_a}$$

from [93], $$V_{2w} = RI_x \sqrt{2(f_3 - f_a)}$$

from [94], $$V_{2T} = \sqrt{8KTR(f_3 - f_a)}$$

The total noise is therefore

$$N_{rms} = \sqrt{V_{1rms}{}^2 + V_{2rms}{}^2 + V_{1w}{}^2 + V_{2w}{}^2 + V_{2T}{}^2} \quad \dots \text{[96]}$$

Considering an example, let $R_1 = 10K$ and $R_2 = 100K$ so that $R = 9K1$. From data for the AD504 amplifier,

$$f_1 = 20\,\text{Hz} \qquad V_x = 9.5\,\text{nV}/\sqrt{\text{Hz}}$$
$$f_2 = 500\,\text{Hz} \qquad I_x = 0.25\,\text{pA}/\sqrt{\text{Hz}}$$

Assume that we wish to compute the total rms noise voltage in the band 1 Hz to 1 kHz, at an ambient temperature of 20 °C. We therefore have the values

$$\left.\begin{array}{l} V_{1rms} = 0.11\,\mu\text{V} \\ V_{2rms} = 0.19\,\mu\text{V} \\ V_{1w} = 0.3\,\mu\text{V} \\ V_{2w} = 0.1\,\mu\text{V} \\ V_{2T} = 0.54\,\mu\text{V} \end{array}\right\} \quad N_{rms} = 0.66\,\mu\text{V}$$

For a crest factor of 3, the noise peaks are likely to exceed $2\,\mu$V for 0.27 per cent of the time. Note that the resistor noise predominates in this example, and indicates the necessity of using low resistance values.

The foregoing calculations provide noise values referred to the input terminals. The output noise in high-gain amplifiers is approximately $G \times$ the input noise, but in low-gain applications, especially where the gain is less than unity, the noise gain will always be at least unity; this is because the equivalent noise-voltage generator appears *at the amplifier input terminals*, and $1/\beta$ is always greater than unity in the inverting configuration. Additionally, due to the phase shifts associated with A and β at high frequencies, there may even be a 'peak-up' of the noise gain at frequencies with low stability margin. A consequence of this is that feedback cannot improve signal-to-noise ratios.

5.4 Popcorn noise

This is a problem which the designer often comes upon by accident. All the rules are obeyed for a low-noise circuit design, and the output noise is then observed on an oscilloscope. The observer sees the general random noise, but suddenly there are large steps of voltage of up to $100 \times$ the random noise, occurring at random times, and lasting for less than 1 ms up to several seconds. This effect is clearly something quite different from the noise discussed in the previous sections. The problem is not mentioned anywhere on the data sheet, and the quoted noise values do not reflect the problem. This is usually because the measuring techniques are designed to detect the average noise power, and do not indicate burst noise. Additionally, the bursts may occur at rates of several hundred a second down to less than one a minute. Information about the measurement of popcorn noise is to be found in reference 17. The cause of this phenomenon is attributable to imperfect semiconductor surface conditions incurred during wafer processing. Manufacturers who specialise in the supply of low-noise amplifiers adopt special fabrication techniques in order to minimise the effect. It is actually the erratic change of bias current between two levels at random intervals flowing in the source resistors which produces the voltage level changes.

It is interesting to note that the proportion of devices exhibiting this problem varies from one batch to the next, and with common amplifiers like the 741, both large and small manufacturers are implicated in supplying poor devices. The manufacturers who supply popcorn noise-free devices tend to do so consistently. Some amplifiers are advertised as being 'popcorn noise free'; for demanding applications it is preferable to select these.

5.5 JFET input amplifiers

With the advent of single-chip amplifiers utilising both FET and bipolar technology, the advantages of JFET input devices and bipolar outputs can be utilised. There are also CMOS op amps, and amplifiers with PMOS inputs and bipolar outputs on a single chip. These high input impedance devices have very low bias currents and input noise currents, and so are of great interest where high source impedance signals are to be processed. MOSFETs generally exhibit greater noise levels than JFETs, but may be of value where the very highest input impedances are required; long-term d.c. bias on MOSFET gates produces a change in the device performance, generally resulting in a drift effect. In JFET devices the noise voltages are caused by channel-resistance Johnson (thermal) noise added to $1/f$ (pink) noise. Current noise is caused by the reverse gate current shot noise. Bipolar amplifier noise does not usually change significantly with temperature, but since shot noise is proportional to the square root of the gate leakage current in JFETs, the current noise outside the flicker region at low frequencies tends to double for every 44 °C increase in temperature (gate leakage current I_{GSS} doubles for about 11 °C temperature increase). Since the noise current is predominantly shot noise, there is very little $1/f$ current noise, and we find that whilst the voltage-noise density spectrum is presented graphically, a specific

value is quoted for current noise (e.g. LF 155, LF 152, 0.01 pA/$\sqrt{\text{Hz}}$)
Multichip hybrid circuits often combine the best of both tech-
nologies, but are invariably more expensive than monolithic devices.

5.6 Noise figure and noise factor

Amplifier input signals will always have noise associated with them,
and so a source data signal-to-noise ratio can be determined. Passing
this combination of signal and noise through the amplifier at a defined
gain is bound to produce a degraded output signal-to-noise ratio due
to the addition of the device noise. Some indication of the magnitude
of this degradation can be obtained by defining *noise factor* as

$$\frac{\text{Input signal-to-noise power}}{\text{Output signal-to-noise power}}$$

Another definition which is often used is

$$\frac{\text{Total available noise power at load}}{\text{Available noise power at load due to thermal noise } R_s}$$

where R_s is the signal source resistance, and the load is at the amplifier
output terminals. The input signal power and input noise power are
both dissipated in the same source resistor, and the output signal
power and output noise power are dissipated in the same load resistor;
it therefore follows that the power ratios can be replaced by the
voltage-squared ratios. The mean square values are finite even though
the mean values are zero, so that we can write

$$\text{Noise factor} = \frac{\overline{V^2_{si}} / \overline{V^2_{Ni}}}{\overline{V^2_{so}} / \overline{V^2_{No}}}$$

where $\overline{V^2_{si}}$ = mean square input signal voltage

$\overline{V^2_{Ni}}$ = mean square input noise voltage

$\overline{V^2_{so}}$ = output signal mean square voltage

$\overline{V^2_{No}}$ = output noise mean square voltage

But $\dfrac{\overline{V^2_{so}}}{\overline{V^2_{si}}}$ = (voltage gain G_1)^2

$$\therefore \text{ Noise factor} = \frac{\overline{V^2_{No}}}{G^2 \overline{V^2_{Ni}}} \qquad \dots [97]$$

The *noise figure* in dB becomes $10 \log_{10}$ (noise factor). Now the output noise is $G \times$ the addition of the source noise and input-referred amplifier noise, so that

$$\overline{V^2}_{No} = G^2 \left(\overline{V^2}_{Ni} + \overline{V^2}_I \right)$$

where $\overline{V^2}_I$ = mean squared amplifier input-referred noise.

Therefore noise figure $= 10 \log_{10} \left[1 + \dfrac{\overline{V^2}_I}{\overline{V^2}_{Ni}} \right]$ ... [98]

Thus if the source is resistive, $\overline{V^2}_{Ni}$ can be replaced by the Nyquist expression for resistance (Johnson or thermal) noise, so that

$$\text{Noise figure} = 10 \log_{10} \left[1 + \dfrac{\overline{V^2}_I}{4KTR_s \Delta F} \right] \quad ... [99]$$

In this way, the noise figure can be used as a figure of merit for comparing amplifier performance when the source resistance and bandwidth are specified. A technique for the measurement of noise figure has been described by Matthews.[18] Noise figure will be zero for an ideal noiseless amplifier; it is important to note that resistors may not produce their calculated thermal noise, and could be worse. Carbon-resistor noise tends to depend on applied voltage.

5.7 Characteristic noise resistance

As the source resistance increases, the amplifier-generated noise increases due to the additional voltage generated by the current noise flowing in the source. There is, therefore, likely to be a value of source resistance where the noise figure, [99], is at a minimum, i.e., where

$\dfrac{\overline{V^2}_I}{\alpha R_s}$ is a maximum, $\alpha = 4KT \Delta F$ being constant

Now $\overline{V^2}_I$ = sum of the input voltage and current produced noise, so that

$$x = \frac{\overline{V^2}}{\alpha R_s} + \frac{\overline{I^2} R_s^2}{\alpha R_s}$$

$$\frac{dx}{dR_s} = -\frac{\overline{V^2}}{\alpha R_s^2} + \frac{\overline{I^2}}{\alpha}$$

which is a maximum when

$$R_s^2 = \frac{\overline{V^2}}{\overline{I^2}} \quad \text{i.e. } R_s = \frac{V_{rms}}{I_{rms}} \quad ... [100]$$

The characteristic noise resistance for minimum noise figure is therefore dependent only on the total amplifier input voltage and current noise. Notice that if the source resistance is made equal to this value, we obtain minimum noise figure, *not* minimum noise; also that optimum R_s may differ from one frequency band to another.

Data presented on noise is often typical; there can be large variations, with the greatest variations occurring in the flicker ($1/f$) noise region, usually less in the white noise-dominated part of the spectrum. Noise can also be dependent on supply, increasing with higher supply voltages. In general, FET input amplifiers will offer improved low-frequency noise performance for applications involving high source resistances, e.g. 1 MΩ and upwards, with bipolar amplifiers producing the best results for lower source resistances.

5.8 Statistical noise reduction

Since the noise of individual devices is uncorrelated, it might be expected that adding together in-phase signal components from several parallel amplifiers would produce an improved signal-to-noise ratio. Figure 5.4 shows a possible arrangement. Let the rms input

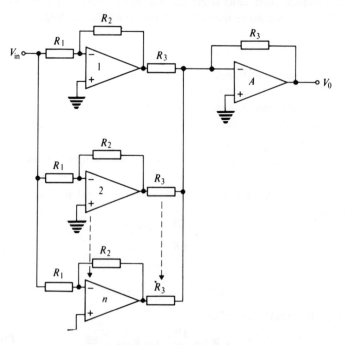

Fig. 5.4 Statistical noise reduction by parallel amplifier operation

signal voltage be S, and the rms input source noise voltage be N. The source signal-to-noise ratio is then S/N. Total input noise to amplifier 1 becomes $\sqrt{N^2 + V^2}$, where V is the rms amplifier noise. The signal output of amplifier 1 is $-GS$ and the noise output is

$$-\frac{1}{\beta} \sqrt{N^2 + V^2}$$

where $G = \dfrac{R_2}{R_1}$ and $\beta = \dfrac{R_1}{R_1 + R_2}$

Signal output from amplifier A is nGS, and noise output is

$$\sqrt{\frac{n}{\beta^2}(N^2 + V^2)} = \sqrt{n}\,\frac{\sqrt{N^2 + V^2}}{\beta}$$

Output signal-to-noise ratio is from amplifier A

$$\frac{nGS\beta}{\sqrt{n}\sqrt{N^2 + V^2}} \qquad \dots [101]$$

The output signal-to-noise ratio from amplifier 1 is $GS\beta/\sqrt{N^2 + V^2}$. Thus the improvement in signal-to-noise ratio from the circuit is $\sqrt{n}$, where n is the number of signal amplifiers.

The above makes the assumption that the noise contributed by amplifier A is negligible, which will be true if R_2/R_1 is high; also that the noise density and bandwidths of each signal amplifier are identical. This approach becomes realistic with the advent of multiple amplifier packages; for example, two quad amp packages could provide $\sqrt{7} = 2.65$ times, or three packages $\sqrt{11} = 3.3$ times improvement in signal-to-noise ratio.

It is reasonable to suppose then, that low-noise transistors could be constructed on the basis of the above reasoning. The LM194 simulates the function of a transistor pair, and is claimed to be 10 to 60 times better than discrete pairs.[19] The construction consists of 100 npn monolithic transistors in two sets of 25 pairs, connected randomly in parallel. Any one pair may have offsets between 0.5 and 3 mV of either polarity, yet when connected in this manner the effective matching is reduced to 50 μ V. This means that voltage drift is reduced to about 0.1 μ V/°C; also good current-gain matching and high common-mode rejection are achieved. The resulting improvement in signal-to-noise ratio of $\sqrt{100} = 10$ times means that device noise in many applications is less than the source thermally generated noise. It is still necessary to consider the effects of the second stage of the composite circuit, since noise and drift in the second stage will be referred to the input divided by the first stage gain. It is therefore necessary to follow the simulated pair by a high-performance amplifier.

5.9 Drift

This term is usually considered to mean input offset voltage and offset current change with temperature, and input offset voltage drift with time. Occasionally a total input-referred drift is specified, and this is expressed by defining a source resistance so that part of the total is produced by the offset current change. The definition of these parameters has been given in Chapter 2, but can be pictorially represented as in Fig. 5.5(a). The voltage error at terminal 1 due to I_{b1} flowing in source resistance R is $I_{b1}R$. The error at terminal 2 is $I_{b2}R$, so that the effective differential error is $R(I_{b2} - I_{b1}) = RI_o$. It is clear that in order to minimise the current-generated error, the source resistances in each terminal should be matched; this is the objective whether inverting, non-inverting or differential configurations are used, if minimum drift is desirable. Having arranged the circuit in this way, it is only offset current and offset voltage I_o and V_o which determine drift. If the differential input is used and a common-mode input voltage applied, it is important that the resistors have tracking

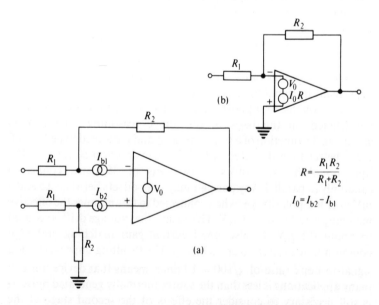

$$R = \frac{R_1 R_2}{R_1 + R_2}$$

$$I_0 = I_{b2} - I_{b1}$$

Fig. 5.5 Equivalent circuit for drift effects

temperature coefficients, otherwise an apparent differential signal will be generated due to mismatch of resistor values, producing the errors discussed in Chapter 4. This error will appear as a temperature-dependent drift at the output, after the initial error has been nulled. The effective circuit can now be drawn as in Fig. 5.5(b). Gain to $V_{in} = -R_2/R_1$. Gain to error voltage is $(R_1 + R_2)/R_1$, i.e. $1/\beta$. Thus in the unity gain situation where $R_2 = R_1$, the gain to the drift voltage generator is in fact 2; this means that the output drift becomes greater than the signal gain times the input referred drift. In low-gain situations, output-stage drift may also become significant, so with fixed-gain instrumentation type amplifiers, drift errors are often specified as an addition of a gain-dependent component and a fixed component.

As a general rule, then, the input *offset* current drift with temperature flowing in the source resistance of one terminal is used to calculate voltage drift for a balanced-differential input, but for single-sided inverting or non-inverting applications where source resistances to each input are not balanced, the input *bias* current drift is used to calculate the effective voltage drift. If the drift is calculated for each terminal separately, they can then be differenced to find the net effect since the sense of the bias current drift is the same for each input terminal.

Bipolar input amplifiers tend to have the lowest input offset voltage drift, with FET input devices having the lowest input offset current drift *at room temperatures*. This last qualification is emphasised, since FET gate currents tend to double for every 11°C increase in temperature, and so it may well be that bipolar input currents for some amplifiers can be lower than FET inputs at high temperatures. Among the techniques used to reduce input errors are:

1. 'Super β' high current gain transistors (β up to 5000, e.g. LM108, AD508).
2. Darlington inputs (e.g. LM316).
3. Temperature compensation of bias current (e.g. LM124, BB3500).
4. Thermally symmetrical geometry (e.g. OP-02, AD510).
5. Laser trimming of on-chip metal-film resistors (LH0044, AD510, AD517, μA714, BB3510).
6. Auto-zero technique (ICL7600 series).

Table 5.3 lists a comparison for some devices between the average input offset voltage drift (typical) and the bias current–produced offset voltage at 25°C in a source resistance of 100KΩ. Chart 5.2 shows curves for variations of total input offset voltage with temperature computed from data for bias currents and average offset

Table 5.3 *Comparison of operational amplifier drift*

	Device	Average offset voltage drift (typ) ($\mu V/°C$)	Bias current at 25°C (typ) (nA)	Offset voltage Produced in 100 kΩ (μV)
FET input	CA3130B	5	0.005	0.5
	BB3521L §	2	0.010	1
	BB3522L §	25	0.001	0.1
	LF155	5	0.015	1.5
	OP-15A	2	0.018	1.8
	TL061	10	0.03	3
	BB3527CM	1	0.002	0.2
CMOS (Auto-zero)	ICL7600	0.005	0.3	30
Bipolar input	LH0044 §	0.2	11	1 100
	LM108	3	0.8	80
	LM216	33‡	0.019	1.9
	AD508K	0.25*	5.5	550
	AD504L	0.3*	42.5	4 250
	OP-02A	2†	15.5	1 550
	OP-07A	0.2	0.55	55
	AD-510L	0.5†	10	1 000
	AD-517L	0.5†	1†	100
	BB3510CM	0.5†	15†	1 500
	SSS725A	0.3	30	3 000
	μA714	0.3	1	100
	LM194‖	0.08 (unnulled)	22 ($I_C = 10 \mu A$)	2 200

* Nulled
† Maximum
‡ Estimated from initial maximum offset at $3.3 \mu V/°C/mV$. Typical effect likely to be less than this
§ Hybrid multichip circuits
‖ Transistor pair

voltage drift figures. It assumes a single-sided input with a source resistance of 100K Ω: the chart will give comparative information only, since offset voltage drift is often non-linear with temperature. The offset voltage error is assumed to be zero at 25 °C, and with constant slope sign; initial bias current – generated voltage is assumed to be nulled along with $V_o S$. The drift performance can be seen to be very much a function of source resistance and temperature. For any given application it is not always clear whether the very low bias current input FET op amps or the very low offset voltage drift bipolar op amps will give the optimum compromise; every requirement must

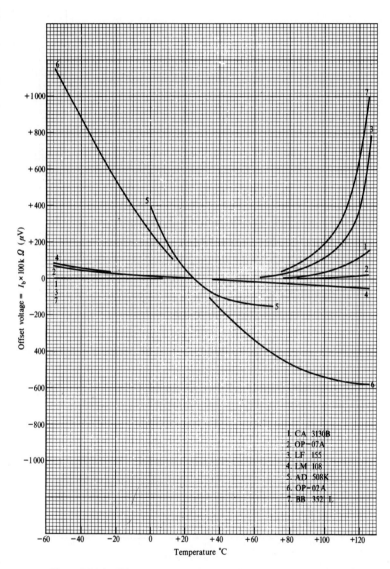

Chart 5.2(a)   Bias current generated offset (unbalanced inputs)

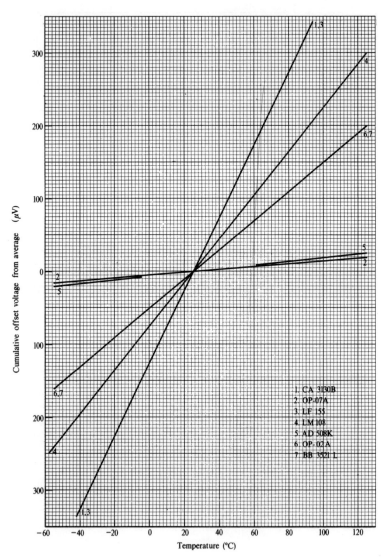

Chart 5.2(b)   Cumulative offset voltage (using average value)

be carefully considered, and the necessary calculations carried out for the operating temperature range required.

Drift with time and 'warm-up' drift are not to be confused. Even though drift with temperature was small, earlier modular-type operational amplifiers tended to have long periods of stabilisation after initial switch-on. This was because the internal dissipation-generated thermal gradients took a long time to stabilise due to the large thermal mass of the modular assembly. This problem is minimised by the good thermally balanced layout of modern IC chips, and the thermal time constants tend to be very. much shorter. Nevertheless, the warm-up time still exists, and may well be exacerbated by thermal feedback from the amplifier output stages if significant power is being dissipated as a result of the load driving requirements. Drift parameters are measured with the devices in a protected environment, so that they are free from heating or cooling sources which may cause thermal gradients; it is imperative that circuit design layout and construction should reflect this precaution if the best drift performance is to be achieved. Thermoelectric e.m.f.s in the input terminals (due to the joining of dissimilar materials) can cause very large temperature-dependent errors if junctions are not maintained at equal temperatures. For example, a typical copper-to-solder input will produce a thermoelectric voltage of about $3 \mu V/°C$ when temperature gradients exist.

If the load driving requirement remains substantially constant or is very small, then device temperature will not change significantly. However, the likely power dissipation and its variations should be calculated, and the resultant chip temperatures computed from the junction-to-case or junction-to-ambient thermal resistances; the likely effects on drift can then be assessed. An early low-drift amplifier ($\mu A727$) utilised on-chip regulated heating, so that chip temperature was held nearly constant ($\pm 3 °C$) for ambient temperature changes between -55 and $+125 °C$. This produced a typical offset voltage drift of $0.6 \mu V/°C$ and an offset current drift of $2 pA/°C$. A maximum of 15 mA heater current was required, and the voltage gain was typically 100, necessitating the connection of another amplifier to provide high open-loop gain. Generally it would be expected that the lower the specified drift performance, the smaller the warm-up drift. It is a rarely quoted parameter, but typical values would be 2–3 min for the AD504, $\mu A714$, OP-07 and LM725; 4–5 min for AD517; or 6–7 min for a 741 to within 1 percent of the final value.

Drift with time implies drift at constant temperature, and is much more difficult to measure and define, simply because of the time scales involved and the variations between devices. We also have the problem of determining whether or not the drift is unidirectional or random in nature. Frequently it is found that a random effect is superimposed upon a continuing trend. Some examples of the presentation of the data are given in Table 5.4. As can be seen, there are such large differences in the methods for defining the value, and in

Table 5.4 *Differences in the methods of drift with time definitions*

Device	Method of drift definition	Typical values
LH0044	Long-term stability	0.3 μV/month
μA777	Input offset voltage drift as a function of time (graphical)	
		100 hours = 32 μV
		200 hrs = 37 μV
		400 = 40
		1000 = 41
BB3500	Average input offset voltage/time	± 2 μV/day
BB3521	Average input offset voltage/time	± 3 μV/day
SSS725	Change in nulled offset voltage. Long-term drift error band (non-cumulative) (graphical)	
		10 hours ± 1 μV
		100 hrs ± 2
		1000 hrs ± 3
OP-07	Total drift with time (graphical)	Random variations of ± 1 μV superimposed on a trend line of: 0.3 μV/month after 2 months 0.2 μV/month after 6 months Worst case 4 μV after 1 month
CA3130/CA3140	Typical incremental offset voltage shift/operating life (graphical)	0.4 mV after 3 000 hours (differential d.c. input = 0 V) 5 mV after 3 000 hours (differential a.c. input = 2 V)
AD504L	Input offset voltage/time	10 μV/month
AD508L	Input offset voltage/time	10 μV/month max + graphical variations between 0 and 192 hrs
BB3510	Input offset voltage/time	0.2 μV/month
μA714	As μA777	0.2 μV/month ultimate

any case the values quoted are usually typical, so that at best they can only be regarded as a rough guide to what can be expected. In many applications, the external components are far more significant in their effects on drift with time, and there are few transducers which can emulate the stability of a good op amp. The MOS input devices CA3130/CA3140 have very large shifts of offset voltage compared with the BB3521 which has JFET inputs and the other amplifiers which are bipolar. The LH0044 and the BB3521 are, however, hybrid circuits.

5.10 Guard shields

In order to exploit the very low bias currents of these devices, care must be taken with printed circuit board layouts, in order to minimise leakage resistance paths. To obtain the best results it may be necessary to hard-wire directly to the amplifier input terminals, but failing this, printed track should surround the input terminals; the track is connected to a low-impedance source at the same level as the signal, to provide an equipotential surface around the high-impedance points. Where the device case is metal and isolated from the device, this should also be connected to the shield; Fig. 5.6 shows the suggested arrangements. Remember that a potential difference of 1 V across an insulation resistance of 1000 M Ω produces 1 nA, a current which is

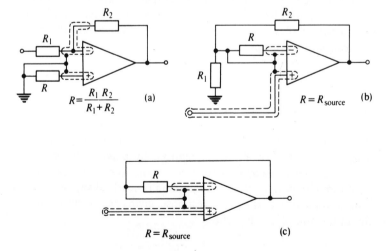

Fig. 5.6 Guard shields for high-impedance circuits.
    (a) Inverting input
    (b) Non-inverting
    (c) Voltage follower

several orders of magnitude greater than the bias current of FET devices. Circuit boards must be protected against environmental contamination, especially in high-humidity environments, either by encapsulation or careful spray-coating with a suitable material.

5.11 Thermal shock

This parameter is usually tested by rapidly changing the device temperature by immersion in a heated silicon oil bath, and a recording is obtained of the output voltage change under specified circuit conditions. These may include overshoot, settling times, or thermal hysteresis, depending on the quality of the amplifier. Good thermal design of the chip minimises these effects, and the performance in the presence of changing temperature gradients is improved. It is unusual for information about thermal shock to be provided by manufacturers on data sheets, but some examples where this information is given are BB3521, AD504, μA714 and BB3510.

5.12 Distortion

There are many ways in which a signal can be changed during its transmission through an amplifier. Some are static by nature, others dynamic. Arguably the most-discussed effects of distortion occur in the design and application of audio amplifiers, about which many texts have been written. The intention here is to mention some of these effects where they are likely to be of particular interest in the application of op amps.

5.12.1 Frequency distortion

Frequency distortion implying the lack of bandwidth and the inability to reproduce different frequency sinusoids at their relative amplitude levels. Feedback is utilised with advantage to extend bandwidth. The shape of complex waveforms is changed by bandwidth limitation, readily demonstrable by slow rise times to transients and sag to pulse tops. Limitation in the frequency domain is nearly always associated with non-linearity in the phase response, giving rise to simultaneous amplitude and phase distortion.

5.12.2 Phase distortion

Phase distortion resulting in time-delay errors for the different frequency components of a complex waveform. The effects have been discussed earlier (Chapter 3). For an absence of phase distortion, a linear phase with frequency response is required.

5.12.3 Non-linearity

Non-linearity is essentially the deviation of the gain from a constant value when the output amplitude changes. It can be expressed in many ways, for example:

1 best straight line;
2 terminal non-linearity;
3 independent non-linearity;
4 percentage slope deviation from ideal.

It is rarely quoted for op amps, but is more frequently available as a specification applicable to instrumentation amplifiers. An example is the LF152, where it is defined as shown in Fig. 5.7.

Gain non-linearity is the curvature of the transfer function from the theoretically perfect function, and is shown in Fig. 5.7 for $G = 1$. It is usual for the magnitude of the non-linearity error to increase with output swing, but is not necessarily always so. The error in the case of the LF152 can be seen to be a constant percentage of reading. Another example is the AD522, shown in Fig. 5.8.

The open-loop gains of op amps are usually very high ($\sim 10^6$), so that for most applications loop gain is high and non-linearity within

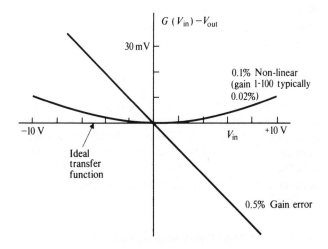

Fig. 5.7 Gain non-linearity
(LF152)

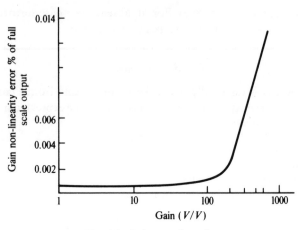

Fig. 5.8 Gain non-linearity
(AD522)

the rated output voltage swing is rarely a problem, since feedback improves linearity distortion.

5.12.4 Crossover distortion

Crossover distortion has been defined in Chapter 2. Again the effects are minimised by utilisation of high loop gain. Specially designed class AB output stages are used (e.g. RC3403A) to remove the effects of crossover in some cases. The RC4136 is one of the very few op amps for which data on distortion is provided; curves are given for total harmonic distortion at 1 kHz against output voltage, and also for distortion against frequency.

5.12.5 Slew rate limiting

Refer to Chapter 2 for a description. Occasionally the slew rate limit is shown to correspond with a defined percentage distortion at the maximum-power bandwidth. When amplifying sinusoids, the frequency can be increased beyond the maximum-power bandwidth providing the amplitude is decreased, for the same distortion content; i.e., the maximum rate of change of output voltage must not exceed the slew rate limit.

5.12.6 Intermodulation distortion

This can be described as an amplitude modulation of a low-level component as a result of a change of operating point, caused by a large-amplitude component at another frequency and operating on a non-linear transfer function. It only occurs when amplifying complex waveforms, and is an important consideration in audio amplifiers. Reference to Fig. 5.9 should clarify the description. When negative feedback is incorrectly used, an amplifier used for music reproduction can sound worse than it does without feedback, even though steady-state measurements or harmonic and intermodulation distortion have been improved. This problem has been shown to be caused by transient intermodulation distortion (TIM).[20] This effect occurs when high values of feedback are applied to an amplifier with low open-loop bandwidth, and is fundamentally an overload situation giving an audible result similar to that caused by crossover distortion. The mechanism of the problem usually occurs where two frequency-compensated amplifiers are connected in cascade, with overall feedback applied. An analysis of the signal levels within the loop shows that voltage limits can be exceeded in response to transient inputs. The reader is referred to reference 20 for a description of the problem and methods for its minimisation. Intermodulation has been used as a method for determining non-linearity. Two sine

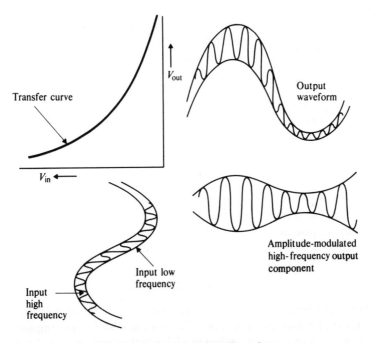

Fig. 5.9 Intermodulation distortion

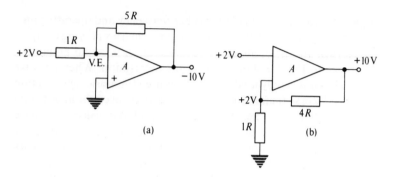

Fig. 5.10 Distortion caused by common-mode voltages

wave signals of differing frequencies are applied to the amplifier, and the subsequent analysis of the sum, difference and other components of the output waveform can be used to compute the non-linearity.

5.12.7 Common-mode effects

If a time-varying low-level signal is amplified by a differential amplifier in the presence of a changing common-mode level, the common-mode gain will contribute a varying error to the output signal, at the rate of change of the common-mode level. The error signal is usually further complicated by the fact that common-mode rejection tends to be a non-linear function of common-mode level and is also frequency-dependent. This effect is readily apparent if we consider the example of Fig. 5.10. Both the inverting (a) and non-inverting (b) circuits have the same gain magnitude. For a.c. or audio applications there is no obvious indication which would be best, assuming input resistance is not a limitation. The loop gain of (b) at $A/5$ is in fact greater than that of (a) at $A/6$, so it might be considered that (b) would have the best performance. It would indeed have a greater bandwidth; however, measurements of output distortion show that (b) is, in fact, inferior. A consideration of the operating levels at the amplifier input terminals shows that (b) operates with a common-mode level of 2 V, whereas (a) is zero. The varying common-mode level of (b) produces the extra distortion.

5.13 Drift-reduction techniques

The dependence of offset voltage and current drift on temperature is often the limiting factor in the precision of a low-level signal amplifier. There is a wide range of options available to the designer for reducing these effects. Many constraints upon the designer normally operate to limit the possible alternatives, such as the ever-present cost/performance compromise, typical or worst-case performance, input noise, required bandwidth, current consumption, etc. This will narrow the choice and help to determine which technique is used. Some of the possibilities are listed below:

1 Integrated chopper amplifier, or auto-zero amplifier.
2 Matched discrete pair and op amp.
3 External drift cancellation techniques.
4 Sample and hold auto-zero correction.
5 Temperature-controlled substrate.
6 Premium op amp, thermally balanced geometry and second-stage drift minimisation.
7 Device selection either by manufacturer or user.
8 Temperature-controlled environment, component ovens.
9 Parametric (varactor) amplifiers.
10 Matched integrated transistor pair plus op amp.
11 Hybrid thin/thick film and chips.

It is clear, then, that the choice is extensive. Usually, the ultimate in low drift performance is supplied by 1 above, but the choice of devices is limited and the cost substantial (discrete amplifiers are not considered here – in some circumstances they can still be a possible alternative). Additionally, chopper amplifiers often have limited bandwidth and higher noise levels than non chopper types. For those interested in pursuing chopper amplifiers and techniques in greater detail, the reader is referred to references 21–30. Chopper amplifiers with differential inputs can be difficult to implement. A recent advance is the commutating auto-zero amplifier (see Chapter 9).

Closely competing would be techniques 5 and 8, but the penalty to be paid here is the increased power supply to provide the heating control and the physical size and weight of the oven (individual metal-can device ovens are available on the market). The solution of 9 has been tried, but the low sensitivity of varactor diodes to the bias voltage and the temperature dependence of the capacitance leads to a low conversion efficiency, and such amplifiers have had limited success.

Very low drift and noise levels can be achieved by matched transistor pairs as in 10, such as LM194, combined with an op amp to provide the high open-loop gain required, but this of course means increased complexity when compared with the device-based solutions of 6 and 7. Selection can be a time-consuming and expensive process and is often not practical in a production environment. This last comment often applies to 3 since it is usually necessary to temperature exercise to measure the drift, and then exercise again to check that the external compensation is providing adequate correction. The premium op amp is often the best compromise, and there are companies who specialise in producing improved versions of standard circuits as well as the proprietary designs.

The auto-zero technique 4 is perhaps best known for its application in analog-to-digital converters and panel meters, and is useful where wide bandwidth is not required. Several methods have been used, and a good circuit has been described by Jaeger and Hellworth.[31] The improved performance is again achieved at the expense of additional complexity, but with the availability of integrated multiplexers, switching problems can be tackled with a minimum parts count.

Some premium op amps are individually trimmed at the chip level to minimise input offset voltage and drift, using either laser trimming of surface-deposited metal-film resistors* or successive shorting of integrated balancing resistors ('zener zap').† In addition, proprietary processes are used in the fabrication of devices in order to minimise leakage currents and increase current gain at low collector currents, e.g., 'triple passivation'.† This can produce initial offset voltages as low as $25\,\mu V$ and drifts of less than $0.5\,\mu V/^\circ C$. One important fundamental technique used to improve thermal stability is the careful design of the chip to achieve thermal balance and minimise thermal gradients. If the output devices are driving load current, dissipation can cause local heating which produces thermal feedback to the input devices, thus producing drift. Graphical data on the LM10 shows the effects of thermal-gradient feedback on offset voltage.

For those readers interested in external drift compensation techniques, see references 32–3.

---

* e.g. Analogue devices, Burr-Brown, Fairchild.
† Precision Monolithics.

6

Feedback – some advantages and limitations

6.1 Review

Feedback as a technique has been with us for a long time. From the very earliest days of valve telephone repeater amplifiers, the advantages of feedback in electronic applications has been exploited. Indeed, as a technique for control, feedback was in use during the eighteenth century as a method for controlling windmills, and closely following these ideas came the well-known Watt-governor used in a closed-loop feedback mechanism to control the speed of steam engines.[34] It is desirable that the student of linear integrated-circuit design should have sufficient background knowledge in order to maximise the benefits to be obtained, but at the same time to appreciate the limitations of feedback.

Numerous techniques have been used to implement amplifiers which are capable of producing very accurate, stable gain performance. Sometimes the methods have been evolved with the principal objective of gain stabilisation, but frequently the improved performance has been the secondary advantage accruing from circuit modifications designed to procure other desirable characteristics.

One of the earliest developments was announced in 1924, and was intended to 'increase the load carrying capacity', 'repeat electrical waves without distortion', and 'suppress the distortion and modulations in an amplifier circuit'.[35]

Most of the network arrangements used over the years can be classified as either 'feedback' or 'feedforward'. Both of these terms are used to describe circuits which attempt to minimise amplifier limitations, but it is not always clear which is the correct term, and occasionally they are used in a completely different context from that attributed by the original authors. An interesting review has been given by Bode,[36] where he describes the evolution of feedback. It is an interesting exercise to attempt to establish when the term 'feedback' was originally used. Black's patent of 1928[35] described an arrangement which is probably more realistically termed 'feedforward', but it is generally considered that he also invented the term 'feedback' in about 1927. However, his formal declaration was not filed until 1932 and eventually appeared as a very comprehensive patent in 1937.[37]

At about the same time, workers at the NV Philips Laboratory in Holland were also active. In a British patent of 1930, 'Improvements in, or relating to, arrangements for amplifying electrical oscillations', the invention was described as follows: 'a portion of the output voltage is coupled back in counteracting sense to the primary side of the amplifier circuit'. This patent was filed in 1928.[38] The amended specification filed in 1929 actually used the word 'feedback'. Soon after this, several publications appeared describing the technique of 'counter-coupling'[39] and 'inverse feedback'.[40-2] Reference 25 attributes the 1928 patent to K. Posthumus, and infers that Black's work appeared at a somewhat later date.

The technique of 'feedforward' should not be confused with 'feedforward compensation'.[43-4] The latter term has been used to describe a frequency-compensation technique for improving bandwidth. Although 'feedforward' was, in fact, the method used by Black, this approach received relatively little attention until more recent times. In 1947 van Zelst described 'feedback' and 'feedforward' without using the latter term, to stabilise amplifier gain.[45] He later went on to describe a gain-stabilisation technique using an auxiliary oscillator system, where the oscillations were non-disturbing to the signal of interest.[46] A considerable volume of work has been done on the application of 'feedforward' to microwave amplifiers to help minimise transit-time effects. One of the earliest applications was described in 1949 as a method 'to reduce distortion without limiting bandwidth'[47], and other, more recent publications have appeared.[48-54] The last reference describes an amplifier arrangement which has improved reliability as a result of the inherent amplifier redundancy, and feedforward has been proposed as a method for improving transient response.[55]

In 1956 a patent was published which described various different ways of applying negative feedback, including secondary loops within the main amplification channel.[56] It is perhaps surprising that so many years had elapsed between the invention of the concept and its application to multiple-loop systems, but it is often the case that theory precedes the technology capable of realising the ideas. More recent circuit configurations designed with the objective of distortion reduction have been described by Sandman using an 'error take-off' principle,[57] and Walker using 'current dumping'.[58] A useful reference to amplifier systems with multistage feedback has been given by Cherry and Hooper.[59]

6.2 The application of feedback

There are a large number of potential advantages accruing from the application of feedback to an amplifier. The desired results are obtained by applying feedback in the correct way. Some factors influenced are:

1. Gain stability.
2. Bandwidth.
3. Transient response/rise time.
4. Input resistance.
5. Output resistance.
6. Distortion.

Note that signal-to-noise ratio is not improved by the application of negative feedback, and can in some circumstances actually degrade it. Examining the possibilities it is clear that there are eight basic alternatives, set out in Table 6.1. Consider the standard inverting op amp configuration Fig. 6.1(a). The gain expression can be derived either from voltage considerations or by simply defining the currents. Irrespective of this, however, the nature of the circuit is the same, i.e., a proportion of the output voltage is fed back in parallel with part of the input voltage and thus is an example of B in Table 6.1. The circuit of Fig. 6.1(b) of the non-inverting configuration is an example of A as is the voltage follower, when $R_2 = 0$ and $R_1 = \infty$. Figure 6.1(c) is an example of series-applied, current-dependent voltage feedback; this connection leads to a whole new family of circuits classified as current

Table 6.1 *Feedback types*

	Series applied	Parallel applied
Voltage feedback proportional to output voltage	A	B
Voltage feedback proportional to output current	C	D
Current feedback proportional to output voltage	E	F
Current feedback proportional to output current	G	H

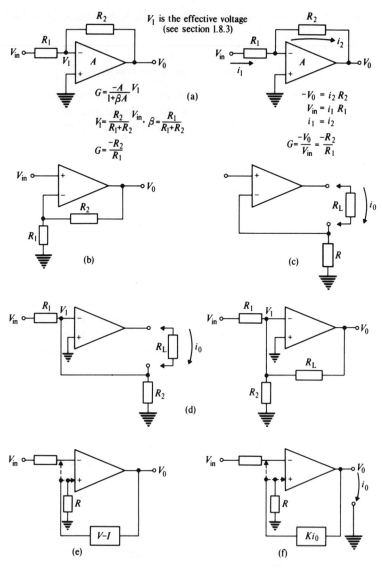

Fig. 6.1 Methods of feedback application

output, which are discussed in more detail later. An interesting case is presented by D in the table, as shown in Fig. 6.1(d).

G, the voltage gain $\approx \dfrac{1}{\beta} \dfrac{V_1}{V_{in}}$ (inverting) (see section 1.8.3) where $\beta = \dfrac{R_1 R_2}{R_L (R_1 + R_2) + R_1 R_2}$ and $V_1 = \dfrac{R_2 R_L}{R_1 (R_2 + R_L) + R_2 R_L}$

Therefore $G = \dfrac{-R_L}{R_1}$

It is clear that the presence of R_2 has little effect on the circuit (providing A is very large), as would be expected, since V_1 has been defined as a 'virtual earth point'.

The other possibilities suggested by the table are shown in Fig. 6.1(e) and (f). Since the operational amplifier is a voltage input device, current feedback must effectively be converted to a voltage at the input terminals, for example, by the inclusion of R. These circuits then revert to one of the other four described above, so do not need further consideration. A fundamental analysis of the possibilities has been given by Pridham.[60] The effects on terminal resistances can be summarised as follows:

Connection	Input resistance	Output resistance
A	Increased	Decreased
B	Decreased	Decreased
C	Increased	Increased
D	Decreased	Increased

The current-equating circuit shown in Fig. 6.2(a) is often useful for analysing computing arrangements where multiple inputs are used.

For this circuit,

$$i_1 = \frac{V_1}{R_1}, \quad i_2 = \frac{V_2}{R_2}, \quad i_3 = \frac{V_3}{R_3}, \quad i_4 = i_1 + i_2 + i_3$$

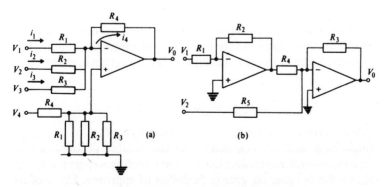

Fig. 6.2 Voltage addition and subtraction

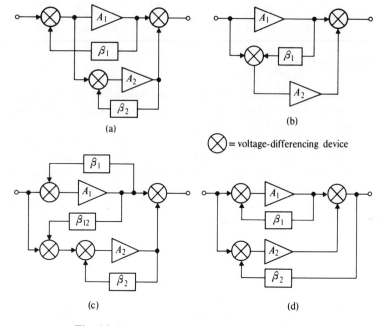

(a)

(b)

$\bigotimes$ = voltage-differencing device

(c)

(d)

Fig. 6.3 Some examples of multi-loop feedback

$$\therefore V_o = -i_4 R_4 + V_4$$

$$= + V_4 - \frac{R_4}{R_1} V_1 - \frac{R_4}{R_2} V_2 - \frac{R_4}{R_3} V_3 \qquad \ldots [102]$$

An alternative method for subtraction is shown in Fig. 6.2(b),

where
$$V_o = \frac{R_2 R_3}{R_1 R_4} V_1 - \frac{R_3}{R_5} V_2 \qquad \ldots [103]$$

For this type of computing network it is often tedious to select the required combination of resistors to minimise the input offset voltages. A technique for achieving this has been given in reference 61.

6.2.1 Multiple-path feedback

In the previous section we have considered the application of a simple single-loop feedback connection. As the number of amplifiers increases, the choice of feedback connections increases enormously, and allows the designer far greater flexibility of approach. Many of the possible arrangements have been documented,[56] but with the availability of multi-op amp packages (e.g. quad amps), it is now much easier to implement the complex networks. Some typical examples are shown in Fig. 6.3, and an example of a gain calculation is given in Appendix 9. Clearly, not all such networks have any significant merit, but in some cases they prove to be useful in achieving higher gain stability without oscillation, reduction of distortion and redundancy in high-reliability circuits.

6.3 Current-source circuits

These circuits can be conveniently classified into two sections, those with the load floating and those with the load referenced to the common line. In some cases current sinking as well as sourcing is required. It is usual to consider a current source as the output resulting from a high voltage connected in series with a very high resistance, as in Fig. 6.4(a). If r_L is small compared with R, then i_0 is constant at V/R irrespective of variations in r_L. The immediate objective, then, is to render the effective value of R as large as possible.

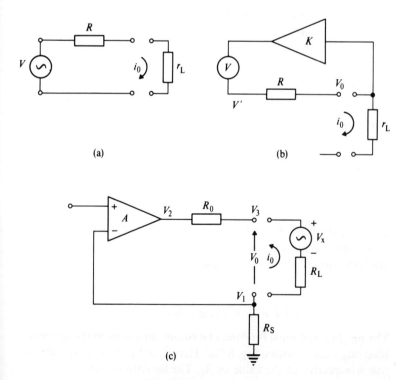

Fig. 6.4 Current output

Again this can be achieved by applying feedback in the correct manner; in fact, the principle is very similar to that of bootstrapping (see section 4.6.3). In Fig. 6.4(b),

$$V' = V + KV_o \quad \therefore \quad i_0 = \frac{V + KV_o - V_o}{R} = \frac{V + V_o(K-1)}{R} \quad \dots [104]$$

and if $K = +1$ then i_0 is constant at V/R.

From [104], $\dfrac{\delta V}{\delta i_0} = \dfrac{R}{K-1} =$ effective output resistance.

Thus the nearer K is to unity, the nearer the effective output resistance is to infinity. The effects of this on a practical circuit can be seen in Fig. 6.4(c). The output resistance as 'seen' by the external voltage generator V_x without the amplifier would be

$$R = R_0 + R_L + R_S$$

$$V_1 = -i_0 R_S$$

$$V_2 = -AV_1 = Ai_0 R_S$$
$$V_3 = V_2 + i_0 R_0 = i_0(R_o + AR_S)$$
$$V_o = V_3 - V_1 = i_0(R_o + R_S[1 + A])$$

Also,

$$V_o = -i_0 R_L + V_x$$
$$\therefore V_x = i_0(R_0 + R_L + R_S[1 + A]) \qquad \dots [105]$$

Considering small changes, effective output resistance is

$$\frac{\delta V_x}{\delta i_0} = R_e = (R_o + R_L + R_S)\left(1 + \frac{R_S}{R_o + R_L + R_S}A\right) \quad \dots [106]$$

$$R_e = R\left(1 + \frac{R_S}{R}A\right) \qquad \dots [107]$$

This is of the form $R_e = R(1 + \beta A)$ where β is the fraction of V_x fed back, and the loop gain βA is the factor providing the apparent multiplication of the output resistance.

6.3.1 Floating-load current sources

The simplest and most fundamental circuit derives from the standard inverting configuration, Fig. 6.5(a). Here $i_0 = V_{in}/R = i_0$, and this is true irrespective of the value of R_L. The limitations are:

1. The source must provide the current.

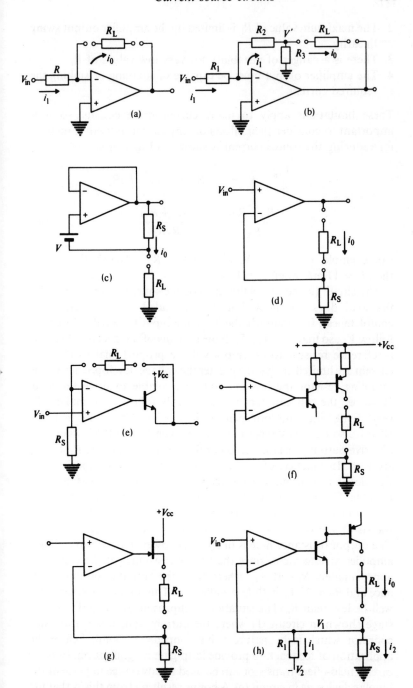

Fig. 6.5 Current outputs for floating loads

2. The maximum value of R_L is limited by the amplifier output swing linearity.
3. There is a danger of instability for very low values of R_L.
4. The amplifier output must be capable of sourcing or sinking the required current.

These limitations apply to many current-drive circuits, so it is important to consider their effects on any circuit utilised. A method for reducing the source current is shown in Fig. 6.5(b).

$$i_1 = \frac{V_{in}}{R_1}; \quad V' = -i_1 R_2 = -(i_0 - i_1)R_3$$

$$\therefore i_0 = i_1 \frac{(R_2 + R_3)}{R_3} = V_{in} \frac{(R_2 + R_3)}{R_1 R_3} \qquad \cdots [108]$$

For example, if $V_{in} = 10\,\text{V}$, $R_1 = 10\,\text{k}\Omega$, $R_2 = 10\,\text{k}\Omega$, $R_3 = 1\,\text{k}\Omega$, then $i_0 = 11\,\text{mA}$ when $i_1 = 1\,\text{mA}$.

The circuit (c) requires a floating voltage source V, and $i_0 = V/R_S$; the load does not float, but it is included in this section for completeness. In circuit (d), the inverting input terminal is forced to follow V_{in}, so that $i_0 = V_{in}/R_S$. Note that any of the circuits which use a reference resistor to develop a voltage proportional to the load current are limited, in that the greater the voltage developed across the sampling resistor the less the voltage available to drive the load. However, the solution does not necessarily follow from making R_S very small, since the loop gain is then reduced resulting in lower effective source resistance. This can be seen from [106], where the effective output resistance becomes $R_e = R_S(1 + A)$, where R_L and R_o are zero. In practice, of course, R_o is always present, and is often significant in determining the magnitude of output resistance obtainable. The higher R_e, the better current source the circuit becomes.

Output current boosting can be provided by the addition of a transistor source as in circuit (e), so that the load current is V_{in}/R_S. The output resistance here is likely to be lower than with the simple amplifier arrangement, since the source resistance at the emitter is approximately $R_T + R_o/h_{fe}$, where R_0 is the amplifier output resistance and $R_T \approx 25/I_e$, with I_e in milliamps. This source resistance may well be less than R_o. The situation is improved by using the output stage shown in circuit (f) where the current source is a transistor collector which is by nature a high-impedance point, before the application of feedback to provide loop-gain magnification. In some cases a field-effect transistor can be used to advantage in place of the bipolar device, as in circuit (g). A bonus resulting from this is that use can be made of the negative gate to source bias of an 'N' channel

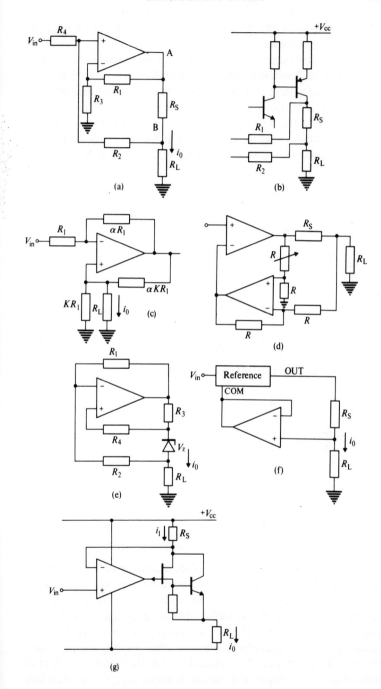

Fig. 6.6 Current outputs for grounded loads

device to maximise the positive excursion at the load before the amplifier output limits.

For many instrumentation and control applications, an output current of 4 to 20 mA is utilised, thus requiring a 'live' zero of 4 mA when the input voltage is zero. An easy way of implementing this is to bias the output as shown in Fig. 6.5 (h). When $V_{in} = 0$, the reference voltage V_1 must also be zero. Therefore $i_2 = 0$ and i_1 is the 'live' zero reference of 4 mA. Thus $R_1 = V_2/4$ mA. If the full range input is V_{in} which corresponds to 20 mA load current, $i_1 + i_2 = 20$ mA, and

$$i_2 = \frac{V_1}{R_S} = \frac{V_{in}}{R_S}$$

$$\therefore i_1 = 20\,\text{mA} - \frac{V_{in}}{R_S}$$

Also,

$$i_1 = \frac{V_{in} + V_2}{R_1}, \quad \text{so that}$$

$$R_S = \frac{V_{in}}{20\,\text{mA} - (V_{in} + V_2)/R_1} \qquad \cdots [109]$$

Substituting in [109] for R_1 gives

$$R_S = \frac{V_{in}}{16\,\text{mA} - 4\,\text{mA}\,V_{in}/V_2} \qquad \cdots [110]$$

For example, if $V_{in} = +1\,\text{V}$, $V_2 = -15\,\text{V}$,

$$R_1 = 3\text{K}75\,\Omega \quad \text{and} \quad R_S = 63\text{R}6\,\Omega$$

An input voltage of 0 to $+1$ V will then produce a load current of 4 to 20 mA.

6.3.2 Grounded-load current sources

The most obvious technique is to place a resistor in series with the load, sample the voltage across it, and use the voltage as a reference for the load current required. This is shown in Fig. 6.6(a). It is assumed that resistors R_1 and R_2 are sufficiently large compared with R_S to cause negligible loading. The circuit actually provides simultaneous negative and positive feedback, but since changes in i_0 will produce a greater voltage change at A compared with B, the overall net effect of the feedback is still negative in nature. A convenient way of analysing circuits is to equate the voltages at the amplifier input terminals; this situation must appertain if the loop gain is high. If R_1 and $R_2 \gg R_L$ and R_S,

$$i_0 R_L \frac{R_4}{R_2 + R_4} + V_{\text{in}} \frac{R_2}{R_2 + R_4} = i_0 \frac{(R_L + R_S)R_3}{R_1 + R_3}$$

If $R_1 = R_2$, $R_3 = R_4$,

$$i_0 \frac{R_4}{R_2 + R_4} R_S = V_{\text{in}} \frac{R_2}{R_2 + R_4}$$

Hence
$$i_0 = \frac{V_{\text{in}} R_2}{R_4 R_S} \qquad \ldots [111]$$

and i_0 is independent of R_L; if $R_2 = R_4$ then $i_0 = V_{\text{in}}/R_S$. The output resistance of such an arrangement is readily calculated with reference to Fig. 6.7.

Let $\dfrac{R_4}{R_2 + R_4} = \dfrac{R_3}{R_1 + R_3} = \beta$, then $V_1 = \beta V_x$ and

$$V_2 = \beta(V_x - i_0 R_S)$$
$$\therefore V_3 = \beta A V_x - \beta A(V_x - i_0 R_S)$$

Hence $V_3 = \beta A i_0 R_S$

$$i_0 = \frac{V_x - V_3}{R_S} = \frac{V_x}{R_S} - \beta A i_0$$

$$\therefore i_0 = \frac{V_x}{R_S} \frac{1}{1 + \beta A}$$

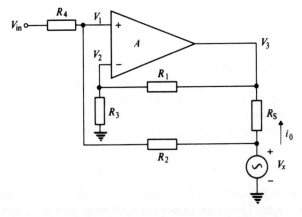

Fig. 6.7 Output resistance of a grounded-load current source

The output resistance is given by

$$\frac{\delta V_x}{\delta i_0}, \quad R_0 = R_S(1 + \beta A) \qquad \ldots [112]$$

It is found in practice that the ideal value of R_0 is rarely achievable; this is largely due to common-mode errors. As the value of R_L increases, the amplifier input terminals 'see' a progressively increasing common-mode voltage; if the resistors R_1, R_2 and R_3, R_4 are not perfectly matched, a significant reduction of the amplifier common-mode rejection capability occurs. (see Chapter 4). The output resistance may be maximised by trimming one of the resistors to optimise the amplifier performance and minimise resistor tolerance effects. Alternatively, the circuit shown in Fig. 6.6(d) may be used where the extra amplifier can be used to eliminate common-mode errors before the feedback is applied to the error-detecting amplifier. The theoretical performance is then more closely approached. The addition of an output stage shown in Fig. 6.6(b) provides higher current sourcing and a much higher output resistance as described for the floating-load circuits. The output current is again V_{in}/R_S.

Figure 6.6(c) shows the well-known 'Howland' circuit, and again exhibits simultaneous positive and negative feedback. An analysis of the circuit is given in Appendix 6. The output current flowing in R_L is shown to be independent of the value of R_L, and equal to $-V_{in}/KR_1$. The current which this circuit can supply is limited, but is nevertheless useful from the merit of its sheer simplicity.

The circuit shown in Fig. 6.6(e) is interesting in that the constant output current flows in the zener reference, the ideal condition for obtaining minimum temperature dependence, providing that the currect zener is chosen for the operating current required. The circuit is, in effect, its own reference, and the output current is given by $i_0 = V_Z/R_3$ provided that $R_1 = R_2$. The analysis is given in Appendix 7. A three-terminal voltage reference is used in Fig. 6.6(f), and a constant voltage is available between COM and OUT terminals. The op amp is used as a voltage follower to maintain the reference voltage across R so that $i_0 = V_{REF}/R_S$. The circuit in Fig. 6.6(g) makes the assumption that $i_1 = i_0$, so that $i_0 = (V_{cc} - V_{in})/R_S$. The FET also acts as a convenient level-shifting device so that V_0 can go to zero.

6.3.3 Frequency stability

Many current-output circuits make use of additional transistors as has been seen. This often causes the open-loop gain to be increased, and the additional devices can introduce high-frequency breakpoints in the open-loop response. When using internally compensated op amps with a low-frequency open-loop roll-off, it rarely causes difficulty, but if wider bandwidth than such devices can provide is required, it is necessary to consider the roll-off and phase shift of the open-loop circuit including the additional devices. If proper consideration is not given to establishing adequate phase margin there is likely to be an instability problem; if the circuit does not actually oscillate, any circuit near to instability will exhibit an increase of noise.

6.3.4 *Constant voltage across a remote variable load*

The circuit in Fig. 6.8 is attributable to Fryer,[62] to whom I am grateful for permission to reproduce it here. It is frequently necessary to provide an excitation voltage to transducers operating remotely from the power supply and the signal-conditioning equipment. In situations where the transducer resistance is variable, the excitation voltage changes, owing to the effects of line resistance. The usual method used to minimise this problem is by the use of an extra wire which senses the load voltage and drives the power supply to restore the line drop. The interesting circuit of Fig. 6.8 can be used without the need for an extra wire in this type of application, providing the line resistance is constant. The maximum variation of load resistance tolerable depends on the capability of the amplifier output to drive the necessary current (clearly if R_L is zero the load voltage cannot be maintained). In practice the value of R_S is adjusted to optimise V_L for variations of R_L; R_C represents the line resistance. The circuit can be considered as neither a constant-current nor a constant-voltage output, but perhaps something between these two limits, since both parameters vary. An analysis of the circuit is given in Appendix 8. The output resistance at X in the circuit is negative, and can be made equal

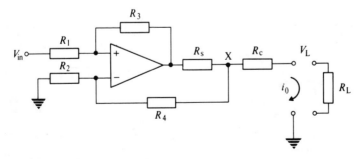

Fig. 6.8 Maintaining constant voltage at a remote variable load

to $-R_C$ by proper selection of components, so that the apparent output resistance at the load is zero.

6.4 Attenuation

The application of op amps to attenuators may seem rather extravagant, but it can often prove to be very useful. Figure 6.9 shows some typical methods. In many cases attenuators can be used in feedback configurations to produce variable gain control. Circuit (a) simply uses a voltage follower to eliminate loading effects on the potentiometer, and (b) uses an inverting amplifier with less than unity gain. It is important to be sure that the open-loop frequency response rolls off at a rate less than $12\,\text{dB}/8^{\text{ve}}$ down to $0\,\text{dB}$ to ensure stability even though the closed-loop gain is less than unity; the loop gain may still be high at frequencies where the loop phase shift approaches $180\,^\circ$. Circuit (b) has a defined input resistance of KR, and both have a very low output resistance. The arrangement in circuit (c) can be useful where the maximum value of usable resistor is limited. Circuit (d) is

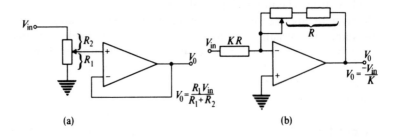

(a)   (b)

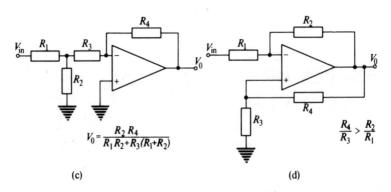

(c)   (d)

Fig. 6.9 Op amps used as attenuators

interesting in that it combines both negative and positive feedback simultaneously. It is necessary to ensure that the level of negative feedback is greater than the positive feedback for the circuit to be stable; hence the bound placed on the resistor ratios. It is useful because accurate high values of attenuation can be achieved by using readily available standard resistor values. For example,

let
$$R_1 = 10K \qquad R_3 = 2K$$
$$R_2 = 1K \qquad R_4 = 1K$$

Then the bound $R_4/R_3 \gg R_2/R_1$ is fulfilled, and

$$V_o = -V_{in} \frac{R_2}{R_1} \frac{1}{[1 - (R_1 + R_2)/R_1][R_3/(R_3 + R_4)]} \qquad \ldots [113]$$

Then

$$V_o = -0.375 \, V_{in} \qquad \ldots [114]$$

6.5 Limitations of feedback and stability margin

We have seen in the foregoing sections how feedback can be applied in various ways in order to achieve a wide variety of circuit functions; it is sometimes interesting to consider the possibilities of one op amp and a few resistors in terms of the many alternative configurations. We have assumed that the feedback is always negative overall, and that the basic frequency or phase compensation is adequate to ensure stability and freedom from oscillation. Circuits which utilise positive feedback are regenerative in nature, and are discussed later. From the list given in section 6.2 it might be assumed that for higher loop gain and increasing feedback, performance is improved (except for, perhaps, transient intermodulation distortion; see section 5.12.6). However, there tends to be a practical limit to how much negative feedback can be applied whilst establishing adequate margins of stability. As the open-loop gain increases, the lower the single-pole breakpoint compensation needs to be to ensure stability at unity gain; this can be seen from Fig. 6.10(a). Amplifiers 1 and 2 have an identical breakpoint f_c, but owing to the higher open-loop gain of 2 the 'safe' roll-off of 20 dB decade is not maintained because of the high frequency dependence of the active devices' characteristics. This suggests that f_c for amplifier 2 may need to be shifted to $f_c/100$ if the gain-bandwidth product of both is to be identical. This can be inconvenient if the breakpoint source resistance is not sufficiently high to enable small capacitors to be used. In addition to this fundamental problem with single breakpoint amplifiers, externally

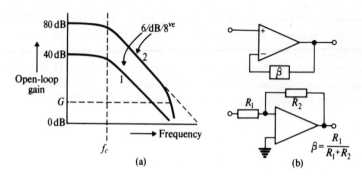

Fig. 6.10 Establishing stable operation

compensated amplifiers are often used when bandwidth maximisation is required. As open-loop gain and bandwidth increases, the effects of stray capacities from physical sources become more apparent, and limit the loop gain achievable. This problem has been discussed by Sandman and one possible solution proposed.[57] It is clearly not permissible to cascade two single breakpoint amplifiers with a response as shown at 1 in order to obtain increased open-loop gain, since the roll-off very rapidly turns over into a $12\,dB/8^{ve}$ slope, thus implying a loop phase shift approaching $180°$. Externally compensated amplifiers can be cascaded, but the resulting combination is difficult to stabilise; it may well be found that the open-loop roll-off must be started at such a low frequency that the loop gain available for wide-bandwidth applications may not be significantly greater than that obtained by using a single amplifier.

Having decided upon the optimised configuration for a specific application, how then do we decide what the stability margin actually is? It may be that the circuit is only conditionally stable, and tolerance variations may cause identical circuits to be unstable. There have been many techniques evolved, but the most frequently used method for op amps is the establishment of 'Bode' plots. These have been briefly mentioned in Section 3.1. Bode established the relationship between attenuation and phase in networks, which is the basis for this approach.[63] He developed idealised characteristics which provide a known phase margin and loop-gain margin within which the amplifier will be stable. A useful résumé of Bode's work has been given by Roddam.[64] The 'ideal' amplifier was assumed to have an asymptotic frequency response roll-off of $10\,dB/8^{ve}$, which led to a constant phase shift of $150°$ and hence a phase margin of $30°$. This form of response leads to a 'peak-up' in the amplitude with frequency plot, and so is rarely used in practice.

With the advent of integrated-circuit op amps and their very high open-loop gain, it is a very simple matter to use a single breakpoint to produce a roll-off of 6 dB/8^{ve}, a constant phase shift of 90°, and hence unconditional stability. However, for externally compensated amplifiers, the 'Bode' plots of open-loop gain with frequency and phase with frequency are useful for determining the phase margin. An example is given in Fig. 6.11. The closed-loop gain is superimposed on the open-loop response; if the loop phase shift at f_c is greater than 90°, then some 'peak-up' can be expected in the closed-loop response. The phase margin can be defined as the extra phase lag required at unity loop gain to cause regeneration and hence instability. It should be noted that the closed-loop gain is termed $1/\beta$; this is valid for the non-inverting op amp configuration, but for the inverting amplifier, closed-loop gain *to the input terminal* is not $1/\beta$, as can be seen from Fig.6.10(b). This should be considered whenever the Bode plots are used to estimate bandwidth or stability.

The 'peak-up' in dB above the low-frequency closed-loop gain can be used as a direct indication of the phase margin, and thus how liable the circuit is to instability. This figure in dB has been called the 'Duerdoth' stability margin. The peak-up in dB considered to be acceptable is arbitrary, but it is generally accepted that $+6$ dB is the maximum tolerable. A method has been evolved whereby the stability margin can be estimated directly from the amplitude and phase

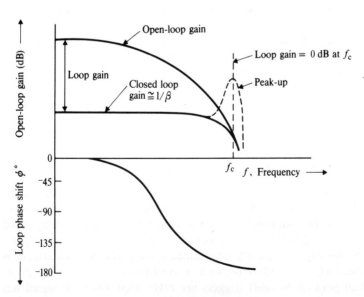

Fig. 6.11 Amplifier 'Bode' plots

plots,[65] and this has been discussed by Edwin and Roddam.[66] The problem can also be solved by use of a calculation chart devised by Felker.[67] The single-breakpoint compensated amplifier is only likely to require such evaluation when the feedback fraction β is not wholly real, but could be a useful approach when multiple-breakpoint amplifiers are being used to maximise bandwidth.

The gain margin can be defined as the ratio in dB between unity and the actual loop gain when the phase margin is zero. Further information on stability criteria is to be found in references 68–71.

6.6 Positive feedback

The use of positive feedback provides us with a very large family of circuits, including comparators, Schmitt triggers, relaxation oscillators, sinusoidal oscillators, etc. They can be designed to have stable limiting states (where the op amp output swings to its own limit or to externally defined bounds), unstable limiting states (e.g. a multivibrator) or to be non-limiting sinusoidal-type oscillators. In addition, there are the wide range of function generating circuits such as sawtooth and triangular wave generators. Many specialised integrated circuits exist which have been specifically designed to fulfil these functions, and optimisation of various parameters such as overload recovery, slew rate, transient response, logic circuit compatibility, etc., renders improved performance compared with op amp realisation of the same functions. However, many general-purpose applications can be realised using op amps at low cost, and manufacturer's manuals can be used to advantage (see for example references 72–4). Useful information is also to be found in references 75–6.

Since many useful circuits are to be found in the literature, it is proposed to outline the basic principles and leave the reader to develop his own circuits for specific requirements.

6.6.1 Comparators and Schmitt triggers

A comparator is a circuit which compares a signal with a fixed reference, and generates a change of state when the signal passes through the reference level. It is thus capable of providing rapid transitions from slowly varying waveforms, and can be used to 'square up' the edges of poorly determined pulses. The Schmitt trigger is effectively a comparator with a controlled degree of hysteresis incorporated. Schmitt triggers are often used when the signal has noise voltages superimposed upon it; the hysteresis is designed to prevent the output from 'toggling' backwards and forwards on the noise, and to provide a single transition when the signal traverses the reference level.

The most elementary comparators simply use to advantage the very high open-loop gain of the op amp, and operate on the principle that a millivolt or less of differential input signal will produce full-scale output. The disadvantages of this are the low speed of response to small signals (especially with internally compensated amplifiers, since the open-loop frequency response applies) and the liability to oscillation during transitions. Overdrive will of course increase the transition speed-up to the slew-rate limit. The transition point is subject to errors due to offset voltage and offset current, which must be nulled for accurate comparison. Some simple examples are shown in Fig. 6.12. V_S and V_R represent the signal and reference voltages respectively, and in (a), (b) and (c) can be interchanged to obtain polarity sensitivity. Care must be taken not to exceed the input differential and common-mode ranges, and the possible non-linearity of the output swing near to saturation should be considered. Zero-crossing detectors are implemented by making $V_R = 0V$, in which case errors generated by the finite common-mode rejection ratio are minimised. It is often found that the positive limit of the amplifier output swing is not the same as the negative limit.

The most frequently used circuits utilise positive feedback to increase the transition speed, provide desired hysteresis and reduce the probability of oscillation during the actual transition. One possibility is shown in Fig. 6.12(d). The magnitude of the hysteresis is determined by the resistor ratios and the amplifier limiting voltages, which may not be of equal magnitude for positive and negative outputs. The rate of transition will be limited by the slewing rate, but parasitic capacitance should be avoided by careful layout and by using the lowest convenient resistor values compatible with current consumption and loading considerations. Resistor R_1 is included for minimisation of offset current – generated offset voltages. Externally balanced amplifiers may be used for their offset-nulling capability. This arrangement is, of course, inverting; the circuit (e) is non-inverting, whilst that at (f) is a variation of (b). Capacitor C will be found useful in speeding up the transition and helping to neutralise capacitance associated with the amplifier input terminal.

6.6.2 *Parasitic hysteresis*

As described above, it is often desirable to introduce hysteresis by intention. However, unrequired hysteresis, or hysteresis greater than the design value, will be experienced and comes as a surprise to the uninitiated. The most easily identifiable problem exists with the simple comparators which have no positive feedback applied. We have seen that the slope of the output change (ideally infinite) is dependent on open-loop gain for small-signal comparison (i.e. where the device is not being over-driven or where the rate of change of input is very slow). Open-loop gain rolls off at very low frequencies for internally compensated amplifiers, so that as the signal frequency increases the open-loop gain, and hence the slope of the output

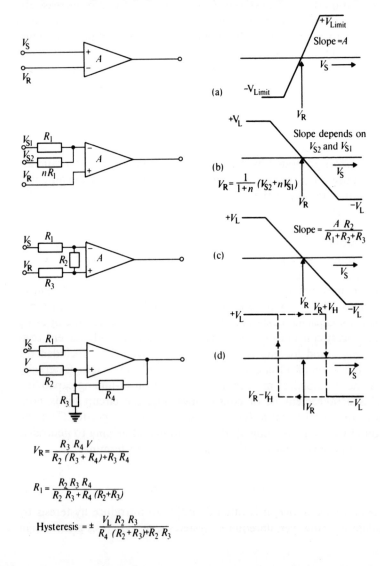

$$V_R = \frac{R_3\,R_4\,V}{R_2\,(R_3 + R_4) + R_3\,R_4}$$

$$R_1 = \frac{R_2\,R_3\,R_4}{R_2\,R_3 + R_4\,(R_2 + R_3)}$$

$$\text{Hysteresis} = \pm\,\frac{V_L\,R_2\,R_3}{R_4\,(R_2 + R_3) + R_2\,R_3}$$

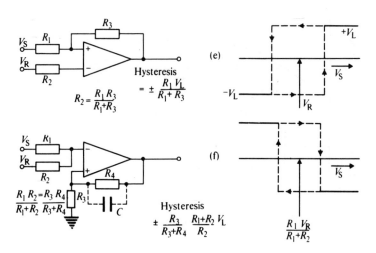

Fig. 6.12 Simple comparators

change, rapidly decreases. The result is that such amplifiers become poor low-level signal comparators at frequencies of a few tens of kilohertz. For example, a typical amplifier may have a gain of 200 000 at 1 Hz but only 100 at 10 kHz. Thus for a 10 kHz signal a 200 mV input change is required for the output signal to swing from -10 to $+10$ V; this, of course, assuming that the full output swing is still available at 10 kHz, a point ascertained by checking the 'full-power bandwidth'. Ideally, of course, we want the output to change states at the reference voltage with the minimum of input voltage change, irrespective of the rate of change of input.

A second source of error which combines with the finite gain problem is that due to phase delay within the amplifier. This occurs even when amplifiers with high slew-rate capability are utilised, and is observable as a time delay between an input voltage change and the occurrence of the corresponding expected output voltage change. Figure 6.13 indicates the nature of the problem. The diagram represents the output of an amplifier comparator without feedback; already hysteresis is present. If feed back is applied, the slope of the output transition increases but the delay is still present. The magnitude of the hysteresis will vary, depending on whether the output slope is frequency-response limited or slew-rate limited. If positive feedback is now applied to such a circuit, there will be little effect on the phase delay, since until the output changes there is no feedback change applied. However, after the delay time a circuit which was originally rise-time limited will speed up until the rate of change of output is slew-rate limited. There will also be an increase of

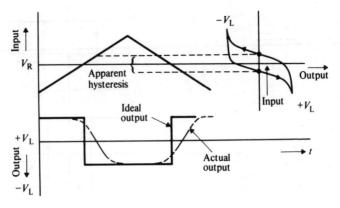

Fig. 6.13 Comparator parasitic hysteresis

hysteresis due to the addition of the feedback-generated 'static' hysteresis.

6.6.3 Output bounding

In the above section we have assumed that the amplifier output is allowed to swing to its saturating limit. Some probable disadvantages of this include non-linearity, increased overload recovery time, unequal positive and negative excursions and increased hysteresis of slew-rate limited, non-feedback comparators. In fact, one method of reducing parasitic hysteresis in the latter case is to limit the output swing. Other reasons for using output bounding include interfacing requirements, waveform clipping, pulse shaping and overdrive limiting to following stages. Some examples are shown in Fig. 6.14. Circuit (a) uses zener diodes to clamp the output to the input of the amplifier when V_o exceeds the zener voltage plus the forward bias diode drop. The zener has capacity and this will degrade the high-frequency performance of the circuit; errors are also incurrred by the leakage current and the slow turn-on characteristic of the zener. A method for improving the performance has been given by Graeme.[75]

Circuits (b) and (d) are basically rectification circuits, but will often prove useful in defining output levels. Circuit (c) is a simple diode clamp, but the active device forms a low source resistance before the diode starts to conduct. Circuit (e) is interesting, not least because of its simplicity. Whilst the input signal or reference are at levels such that the diode is reverse biased, the output follows the input if the amplifier input resistance is high and bias-current errors are small. As

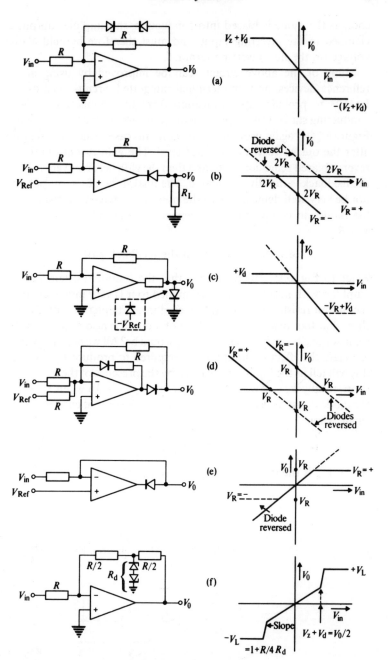

Fig. 6.14 Output bounding

soon as the diode is biased into conduction, the amplifier output is clamped to the inverting input, and must therefore be held at the voltage on the non-inverting terminal.

Many of the above circuits can be improved by using active reference devices, and two-terminal integrated circuits such as the LM103 to provide rapid transition from the conducting to non-conducting states, thus improving the sharpness of the 'knee' voltage. Figure 6.14(f) again uses zener diodes, in this case to increase the gain after the output reaches a defined level. The gain is unity until the reverse-biased zener conducts, and then the gain becomes very high as the feedback is rapidly attenuated. The final slope of the output will not be very well defined due to the non-linear dynamic resistance of the diodes in conduction, and depends on the ratio between $R/2$ and R_d.

6.7 Non-linearity in the feedback path

Refer to Fig. 6.15. We have seen that the output resistance of an amplifier with negative feedback applied to it is very much less than the output resistance of the amplifier alone. (Chapter 2). Could it be, then, that the forward voltage drop of a diode is also reduced in the same way as its intrinsic resistance R_d would be? All we need to do is to ascertain the effective input voltage required to produce $V_o = 0$, and this will tell us the input error caused by the presence of the diode volt drop (ideally $V_o = 0$ when $V_{in} = 0$). Now $V_2 = +0.6$ when $V_o = 0$ if the diode is just biased into conduction, but the output current is very small. Then

$$V_1 = \frac{V_2}{A} = V_{in}\frac{R_2}{R_1 + R_2}$$

where V_{in} is the input required to produce $V_o = 0$.

$$\therefore V_{in} = \left(1 + \frac{R_1}{R_2}\right)\frac{0.6}{A} \quad \text{for } V_o = 0$$

$$= \frac{0.6}{A}\frac{1}{R_2/(R_1 + R_2)} = \frac{0.6}{A}\frac{1}{\beta R_2/R_1} \quad \dots [115]$$

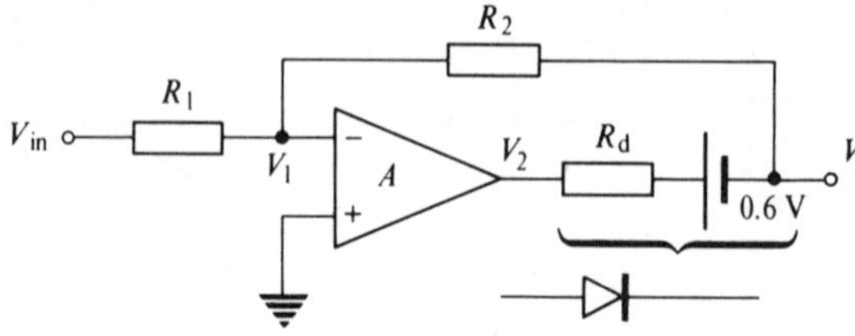

Fig. 6.15 Non-linear feedback

where
$$\beta = \frac{R_1}{R_1 + R_2}$$

$$\therefore V_{in} = \frac{0.6}{\beta A} \frac{R_1}{R_2} \qquad \ldots [116]$$

For the unity gain situation where $R_1 = R_2$, the input-referred error due to the diode voltage drop is $0.6/\beta A$, i.e., the drop divided by the loop gain. It therefore follows that changes of V_d and hence the effects of temperature are also reduced by βA, and the value of the active rectifying circuit is readily apparent. For an input signal of V_{in}', the output would be

$$V_o = \frac{R_2}{R_1} \left(V_{in}' + \frac{0.6}{\beta A} \frac{R_1}{R_2} \right)$$

$$= \frac{R_2}{R_1} V_{in}' + \frac{0.6}{\beta A} \qquad \ldots [117]$$

As the closed-loop gain increases, β decreases, so that the error signal increases as would be expected.

The above example indicates how powerful negative feed back can be in reducing the effects of non-linear components within the loop, and leads to a large family of functional circuits for curve generation, non-linearity correction, and clipping or clamping techniques. The advantages can be seen, for example, in Fig. 6.14 (b), (d) and (e).

6.8 Oscillators

If a proportion of an amplifier output signal is connected to the non-inverting input, the loop can become unstable and the output will saturate in one direction. If this feedback is then switched in sign or is limited so that saturation does not occur, a free-running oscillatory circuit may be realised. Alternatively, if the feedback signal is phase shifted by $180°$, then connection to the inverting input will produce a loop phase shift of $360°$ and again an oscillator is produced. Unstable circuits with positive feedback will generally oscillate so that the output stages limit unless automatic gain control is applied to prevent saturation; sinusoidal oscillators require some form of gain limiting to obtain a low-distortion waveform. If a high-gain amplifier has been inadequately frequency-stabilised, it will sometimes oscillate at a very high frequency (i.e. where the loop gain is unity) and produce an almost sinusoidal output. In these cases, some form of amplifier limiting is operating automatically to prevent output saturation. The output of a high-gain d.c. amplifier may occasionally appear noisy or unstable when it is displayed on a DVM or other averaging device; this can often be traced to a high-frequency circuit instability. The rule is clear: even when building a low-frequency amplifier it is important to check the output noise level with an oscilloscope. Radio transmission can also cause difficulty in this way, with the BBC transmissions on 200 kHz in the long-wave band (1500 m) being a frequent source of trouble in the UK. In this section, however, we consider some of the possible alternatives in the realisation of specific oscillator requirements. Op-amp based oscillators can be conveniently separated into two groups: saturating and non-saturating. The former term implies that the device output is driven to a limiting condition, whether as a result of internal saturation and cut-off or by external bounds. The non-saturating circuits probably form the largest group of circuits, but it is possible that the saturating types are the most frequently utilised.

6.8.1 Saturating oscillators

Astable, monostable and bistable multivibrators (i.e. unstable, having one stable state and having two stable states respectively) are the most familiar circuits, although it is engineering licence to call a bistable circuit an 'oscillator'. However, the relevance of the grouping stems from the fact that the circuit configuration is similar and only small changes are required to produce one from the other.

6.8.1.1 Astable multivibrators. Figure 6.16(a) shows the basic astable circuit. Assume at switch-on the capacitor has zero charge and that the output arbitrarily takes up the positive output limit; the positive feedback via R_1 and R_2 speeds up the output transition to one of the limiting states. C charges towards V_o on a time constant of CR_3. When the voltage on C reaches that of the positive input terminal which is held at $+V_o R_1/(R_1 + R_2)$, the output begins to change (subject, of course, to the input offset error). Regeneration occurs as a result of positive feedback, and the output rapidly switches to its negative output limit. This causes the current through R_3 to flow in the opposite direction, and the charge on C changes accordingly. The positive input terminal is now held at $-V_o R_1/(R_1 + R_2)$, and C charges negatively until this new threshold is reached, when the output changes state again. Notice that only one capacitor is required, whereas with a discrete transistor astable, two are usually needed, and that the speed of output transition depends on the amplifier slew-rate capability, and not on time required to change the charge on C. Notice also that T_1 will not necessarily be identical to T_2, since $+V_o$ is not

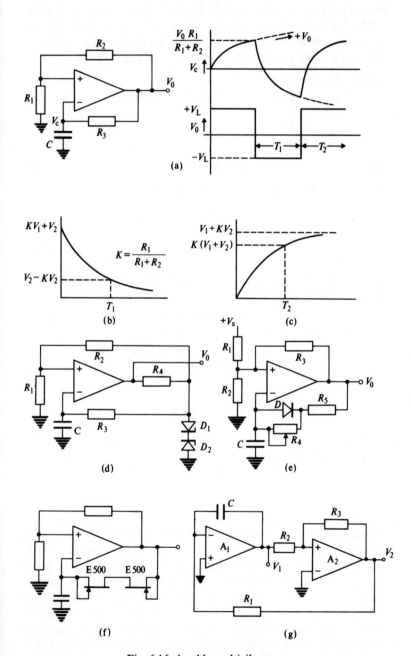

Fig. 6.16 Astable multivibrators

always the same as $-V_o$, and the levels set by R_1 and R_2 will differ.

The simplest way to calculate the time T_1 is to shift the zero reference and assume that C is discharging from a total voltage of $KV_1 + V_2$ down to the point $V_2(1 - K)$ (see Fig. 6.16(b))

where $\quad V_1 = +V_o$
and $\quad V_2 = -V_o$ $\Big\}$ $\qquad$ and write magnitudes

$$K = \frac{R_1}{R_1 + R_2}$$

Then $\qquad\qquad V_c = (KV_1 + V_2)e^{-T/CR_3}$ $\qquad$ ... [118]

where V_c is the voltage across C at any time T.

$$\therefore \text{ at } T_1, \quad V_2(1 - K) = (KV_1 + V_2)e^{-T_1/CR_3}$$

$$\therefore T_1 = CR_3 \log_e \frac{V_2 + KV_1}{V_2(1 - K)} \qquad \text{... [119]}$$

To calculate T_2, we can use Fig. 6.16(c) and consider the effective charging expression

$$V_c = (V_1 + KV_2)(1 - e^{-T/CR_3}) \qquad \text{... [120]}$$

$$\therefore K(V_1 + V_2) = (V_1 + KV_2)(1 - e^{-T_2/CR_3})$$

$$\therefore T_2 = CR_3 \log_e \frac{V_1 + KV_2}{V_1(1 - K)} \qquad \text{... [121]}$$

If $V_1 = V_2$, then $\qquad T_1 = T_2 = CR_3 \log_e \frac{1 + K}{1 - K}$

Since both T_1 and T_2 are directly proportional to R_3, the frequency can be adjusted by changing R_3. Improved amplitude accuracy can be achieved by clamping the output using one of the arrangements shown in Fig. 6.14. The speed is limited by the amplifier slew rate, so that selection of an amplifier with higher slew capability will enhance the maximum frequency obtainable. When the amplifier is operating with significant slewing limitation, the periods T_1 and T_2 are no longer accurately defined by [119] and [121], since the capacitor will be changing its charge during the time taken for a transition. The circuit will operate at lower frequencies with specified values of C and R_3 than the transistor multivibrator since a larger proportion of the output swing can be used by defining the comparision level with R_1 and R_2. The symmetry of the output waveform may be varied by returning R_1 to a controlled voltage other than ground. Errors will be incurred by increasing R_3 to such a magnitude that input bias current affects the voltage on V_C. Advantage may be taken of the MOS or

JFET input op amps with their very low bias currents. The circuit will always 'self start' (often a problem with discrete multivibrators) since the d.c. negative feedback is always greater than the positive feedback, thus preventing a 'latched' situation.

The circuit of Fig. 6.16 (d) uses R_4, D_1 and D_2 to clamp the feedback levels. This stabilises the frequency against variations of power supply voltage and temperature dependence of amplifier output swing. Single-supply circuits can be conveniently realised by using one of the op amps with a common-mode input voltage range of 0 V minimum. Figure 6.16 (e) shows an example of a single-supply circuit. Resistors R_1 and R_2 are selected to bias the input away from zero when an amplifier without 0 V common-mode limit is utilised, and R_4 in this case is used to provide a variable repetition rate constant-width output pulse. With D connected as shown, the negative-going output is of constant width since only R_5 determines the charging time of C. Connecting the diode in the other direction keeps the positive-going output at constant width.

A quasi-triangular waveform is available at V_C in circuit (a). The smaller the value of K, the more triangular the waveform, since only a small part of the exponential curve is utilised. If R_3 is replaced with a bidirectional current source, then the output at V_C will be a good triangular waveform since C charges and discharges with a constant current. The times T_1 and T_2 will then be given by

$$T = \frac{KC(V_1 + V_2)}{i} \qquad \ldots [122]$$

where i is the magnitude of the current source. FET transistors can be connected as bidirectional sources,[12] but some care must be exercised in the control of voltage swings to ensure that the devices always operate in the 'pinch-off' mode. Another alternative is to use current-limiting diodes (actually two terminal devices based on an FET with internal gate connection) as shown in circuit (f). Selection of different devices then determines the current i in [122]. Improved control of square wave and triangular wave generation can be achieved by using the two-amplifier circuit shown in (g). It is basically an integrator A_1 followed by a comparator A_2 with hysteresis. The slope of the triangular waveform at V_1 is set by CR_1 and the period by the ratio of $R_2/(R_2 + R_3)$. The hysteresis voltage of the comparator is $\pm V_2R_2/(R_2 + R_3)$, and the output of the integrator ramps up or down until the trigger threshold is reached. The output state of A_2 then changes over and the ramp proceeds in the opposite direction until the other threshold is crossed. For accurate output levels offset adjustment can be provided to A_2 – input and for a symmetrical waveform an adjustable voltage is applied to A_1 + input.

6.8.1.2 Monostable multivibrators.

With the advent of TTL and CMOS digital integrated circuits, the necessity for using op amps to produce monostable and bistable functions dwindled. However, these circuits can still be useful occasionally, and they are included here for completeness. The potential advantages could apply to multi-amplifier packages where pulse circuits and linear circuits can be implemented simultaneously within a single package. A simple 'mono' can be implemented as shown in Fig. 6.17(a). At switch-on, the positive bias applied to the non-inverting terminal drives the output to its positive limit; the change of level is transferred by C_2 to the input and the positive feedback speeds the transition. Diode D clamps the input at $+V_d$. The negative triggering transition is transferred by the small capacitor C_1 and causes the output to go negative, again aided by the positive feedback loop. The charge on C_2 must change, and the effect is shown in the waveform sketch at in Fig. 6.17(b). The period is determined by the exponential part of the waveform V_A, and can be described by the equivalent circuit in Fig. 6.17(c). The voltage at V_A is given by

$$V_A = \left[V_c + \frac{KR_1}{R+R_1}(V_1 + V_2) - V_d \right]\left[1 - e^{-T/C_2(R+R_1)} \right] \quad \ldots [123]$$

at $T = T_A$, $\quad V_A = \dfrac{KR_1}{R+R_1}(V_1 + V_2) - V_d, \quad V_1 = +V_o, \quad V_2 = -V_o$

$$\therefore \ T_A = C_2(R+R_1)\log_e\left[1 + \frac{KR_1}{R+R_1} \frac{(V_1+V_2) - V_d}{V_c} \frac{}{V_c} \right] \quad \ldots [124]$$

If $R \ll R_1$, and $V_2 = V_1 = V_c$,

$$T_A \approx C_2 R_1 \log_e (1 + 2K) \quad \ldots [125]$$

if $K = \frac{1}{2}$,

$$T_A \approx 0.69\, C_2 R_1 \quad \ldots [126]$$

The period of T_A can be controlled by varying the value of V_c.

An alternative circuit is shown in Fig. 6.17(d) and the associated waveforms at (e). At switch-on C_1 is a low lmpedance and V_A is zero; if the negative output state is taken up it will be transitory and the quiescent state is V_o at the positive limit. V_A is then clamped at $+V_d$, and V_B will always be more positive at KV_2. A negative trigger pulse (C_2 is small) takes V_B and hence V_o negative, with the positive feedback loop speeding the transition. C_1 then charges negatively through R_1 towards V_2 until it reaches V_B, which is held at KV_2. The output then changes state and C_1 recovers to $+ V_{d1}$ towards V_1 on the same time constant. This recovery time limits the re-triggering capability of the circuit, but this can be improved by connecting R_4 and D_2 as shown and making R_4 much less than R_1. Assuming that the exponential part of V_A starts at 0 V,

$$KV_2 = V_2(1 - e^{-T_A/C_1R_1}) \qquad \ldots [127]$$

then
$$T_A = C_1R_1 \log_e \frac{1}{1 - K} \qquad \ldots [128]$$

A variant of the circuit (a) can be seen in Fig. 6.17(f). We have the positive feedback loop via C_2, but this time the negative terminal has a negative bias applied through R_1. This holds the output positive until a positive trigger sufficient to take V_A above zero takes the output negative. The period is then determined by the time taken for V_B to rise from $- (V_1 + V_2)$ to V_c towards 0 V.

Then
$$T_A = C_2R_2 \log_e \frac{V_1 + V_2}{V_c} \qquad \ldots [129]$$

The period can be controlled by varying V_c or returning R_2 to a potential other than zero.

The circuit in Fig. 6.17(g) is a modification of that shown at (d). We again have the positive feedback loop established by R_2 and R_3 with the trigger applied to the positive input. This circuit needs to be carefully biased for correct functioning. Initial conditions are:

$$V_A = V_1 - (V_Z + V_D) \qquad \ldots [130]$$
$$V_B = KV_1 + (1 - K)V_c \qquad \ldots [131]$$

For correct operation, V_B should be less than V_A. After triggering, V_D becomes reverse-biased and C_1 discharges through R_1 until $V_A = V_B$.

Then
$$T_A = C_1R_1 \log_e \frac{V_1 - (V_Z + V_D)}{(1 - K)V_c - KV_2} \qquad \ldots [132]$$

The circuits described above have been shown in configurations producing negative-going pulses; but several of them can be used for positive outputs. For example, if in Fig. 6.17(a) V_c is returned to a negative voltage and D is reversed, a positive-going trigger will produce a positive output pulse. Again in Fig. 6.17(f), returning V_c to a positive bias forces the output initially to its negative output limit, and a negative trigger pulse will produce a positive-going output. It is important to note that all of the expressions for pulse width will be

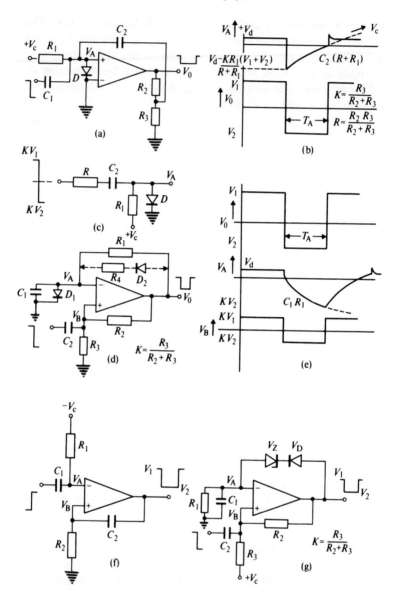

Fig. 6.17 Monostable multivibrators (see text for positive output pulses).

approximate, owing to component tolerances, input offset voltages and poorly determined output swings. In order to define V_1 and V_2 more precisely, an output clamp such as that shown in Fig. 6.16(d) may be used.

A very useful circuit is that shown in Fig. 6.18(a). It requires two amplifiers but serves as a comparator and a mono producing simultaneous positive and negative output pulses. The operating waveforms are shown in Fig. 6.18(b). The bias voltage V_c forces the initial conditions of V_{o1} high and V_{o2} low. R_1 and R_2 provide positive feedback around the loop for latching operation. A positive input trigger takes V_A above V_c and causes V_{o1} to go negative. Since the

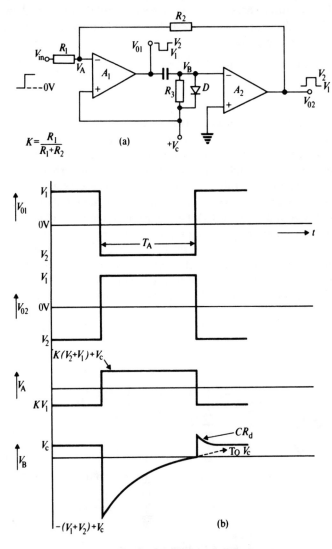

Fig. 6.18 Comparator and monostable circuit

input is d.c.-level sensitive, correct biasing and selection of R_1 and R_2 provide comparator action. The threshold voltage will be $V_{in} = (V_c + KV_1)/(1 - K)$ (writing magnitudes for V_1). At this input level V_{o1} falls, V_B falls from V_c to $-(V_1 + V_2) + V_c$, V_{o2} rises and the feedback speeds the transition and takes V_A up to

$$V_A = KV_2 + (1 - K)V_{in} = K(V_2 + V_1) + V_c \quad \ldots [133]$$

writing magnitudes for V_1 and V_2. V_B rises towards V_c until it reaches 0 V, then the input threshold of A_2 is reached and the outputs change state. Diode D clamps V_B and prevents it from following V_{o1}, quickly changing the charge on C.

Pulse width $T_A = CR_3 \log_e \dfrac{V_1 + V_2}{V_c}$ $\quad \ldots [134]$

6.8.1.3 Bistable multivibrators. It is a very simple process to implement a bistable circuit. All that is required is a positive feedback path to speed the transition and establish a latching state at the output, and a triggering signal applied in such a way as to disturb the stable condition. This can be fulfilled by the circuit in Fig. 6.19(a). The ratio of R_2 to R_3 establishes the level at the non-inverting input defined by the output bound $\pm (V_Z + V_D)$. The trigger input must be of sufficient magnitude to take the inverting input V_A greater than V_B to change the output state. Unfortunately, the first change of state will occur on a positive-going edge but the subsequent change on a negative edge. This problem could be removed by the circuit shown in Fig. 6.19(b), although the increased complexity will seldom be merited. When the

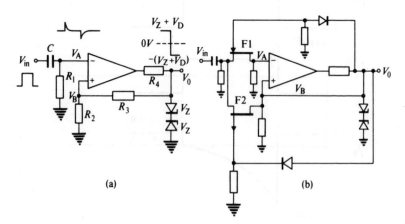

Fig. 6.19 Bistable circuits

output is high, V_B is high and V_A must be increased to trigger the circuit. Under these conditions F_1 is on and F_2 is off. The trigger pulse therefore raises V_A. When the output is low, F_1 is off. The next positive trigger pulse is therefore applied to V_B which is raised above V_A (at 0 V) and causes the output to change state again.

6.8.1.4 Voltage-controlled oscillator. The circuit of Fig. 6.20(a) is based on that of Fig. 6.16(g). When the output state is low, T_R is off and D protects the reverse V_{be} rating of the transistor. Current i flows through R_1 and the output of the integrator ramps downwards. When the ramp crosses the comparator threshold (set by the hysteresis produced by the feedback components), the output goes high and switches on T_R. Current $2i$ now flows from the source through R_2, as the resistors have been chosen to make the transistor current twice the integrator current. This causes the output ramp to reverse direction with an equal but opposite slope. The ramp will then pass back through the comparator threshold and change the output state again. Figure 6.20(b) shows the expected waveforms. If the control voltage V_c is changed, the input currents change and alter the slope of the

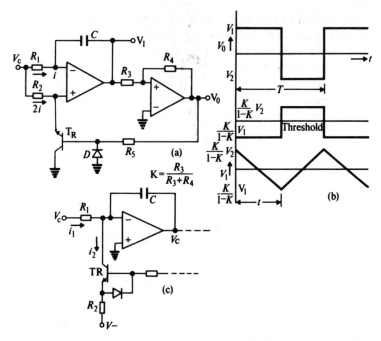

Fig. 6.20 Voltage-controlled oscillator

ramp thus changing the frequency. The frequency is determined by the input current and the threshold voltages.

$$i = \frac{V_c}{R_1}. \quad \text{Ramp slope} = \frac{i}{C}$$

$$\therefore T = \frac{\dfrac{K}{1-K}(V_1 + V_2)C}{i} \qquad \ldots [135]$$

Hence period $T = \dfrac{2CK(V_1 + V_2)R_1}{(1-K)V_c}$

$$\therefore f = V_c \times \frac{(1-K)}{2CKR_1(V_1 + V_2)} \qquad \ldots [136]$$

An alternative method for implementing this circuit is shown in Fig. 6.20(c). With this arrangement the positive output state turns T_R on and $i_2 = 2i_1$ when R_2 is carefully selected. This has the effect of removing charge from C at the same rate as during the charging process, since the current flowing in the feedback loop is i_1 in both cases. The frequency of operation can be derived in the same way as in [136].

It is interesting to note that in the circuit (a) the transistor operates as a switch, so that R_2 is connected to ground or otherwise, whereas at (c) it is designed to be a constant current sink. In the case of the switch (a), the turn-on base current must always be greater than $2i$ for the switch to have a low resistance. It will sometimes be found that transistors can be operated with improved switching performance by forward-biasing the base-collector junction as shown in the inset; alternatively, a suitably biased FET could also be used.

A very simple method of producing a pulse-width modulated signal is to apply a triangular wave such as that derived from V_1 of

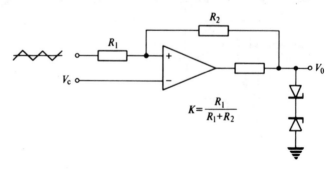

Fig. 6.21 Pulse-width modulator

Fig. 6.16(g), to a comparator whose input threshold is varied according to the control voltage. Figure 6.21 shows the arrangement. The back-to-back zeners render the output independent of supply voltage variations, and R_1 and R_2 establish the hysteresis band. As the threshold voltage V_c is changed, the time taken for the triangular input waveform to cross the threshold varies, thus modulating the width of the output pulse. The transition points on the input waveform are given by

$$V_t = \pm \frac{(V_c - KV_o)}{1 - K}$$

6.8.2 *Non-saturating oscillators*

Circuits with this classification are generally assumed to be sinusoidal oscillators, although waveform generators (6.9) may well be non-saturating in nature. Essentially a sinusoidal oscillator is produced by connecting sufficient positive feedback around a loop so that instability is maintained, but by limiting that feedback dynamically so that the amplitude does not continue to increase until non-linearities prevent its further growth. In other words, the loop gain is maintained precisely at unity. There are numerous techniques for producing the desired phase shift and controlled gain conditions, and a few of the alternatives are discussed below. Among the techniques for determining the required frequency of oscillation are various C–R passive networks, tuned-circuit L–C networks, crystals and ceramic devices. For amplitude control, temperature-dependent devices such as thermistors and lamps can be used, or diodes, transistors and FETs as gain-control devices. The decision about which amplitude control method is used depends on such considerations as the tolerable harmonic distortion in the output, the speed of response to changes, and the dependence on ambient temperature. Many of the traditional tuned-circuit oscillator arrangements may be used, e.g., Clapp, Colpitts and Hartley, but they tend to be less used in present times due partly to the incompatibility of inductors with integrated circuits, and partly due to the more sophisticated integrated techniques available today (e.g., frequency synthesis and phase-locked loops). Inductors are often utilised in the oscillator circuits of switch mode power supplies. For those readers interested in high-frequency transistor oscillators using tuned circuits, reference 77 is recommended.

6.8.2.1 R–C oscillators (phase shift). One of the most frequently used circuits is that based on the 'Wien bridge' phase-shifting network. An analysis of this is given in Appendix 10. The simplest circuit is that of

Fig. 6.22(a) and at the frequency $f_n = 1/2\pi CR$ there is a minimum phase shift around the positive feedback loop and the attenuation is $\frac{1}{3}$. When the amplifier loop gain is $\times 3$ the circuit will oscillate at f_n. It starts oscillating as a result of noise which is readily filtered to provide the cumulative effect at the minimum phase frequency. The negative feedback loop limits the loop gain to the value required to just maintain oscillation, but to prevent the output from saturating and thus introducing distortion. This can be done by selecting a suitable miniature lamp such that at the maximum permitted output voltage swing the resistance ratio of $R_L/(R_L + R_2) = \frac{1}{3}$. If the voltage increases beyond this value, the increased current heats the filament further, the resistance increases and the feedback signal increases, thus decreasing the gain and maintaining the desired gain control. Alternatively, the lamp is replaced by R_1 and thermistor $-t$ replaces R_2. The thermistor is a directly heated negative coefficient type, and this time self-heating of the thermistor reduces R_t and increases the feedback to decrease the gain. These rudimentary circuits are not easy to implement due to the difficulty of determining adequate drive current, and they also suffer from long time delays and dependence on ambient temperature.

Various improved techniques are possible, and many circuits utilise FETs as voltage-dependent resistors to work as feedback attenuators. Some examples are shown in Fig 6.22(b)–(d). The principle in each case is the same: the output signal is detected (rectified) and smoothed to produce a d.c. level which is used to control the resistance of the FET in such a way that the feedback factor becomes $\frac{1}{3}$ and establishes a closed-loop gain of 3. The components and configuration used largely depend on the characteristics of the FET chosen. The C–R circuit is connected to the non-inverting terminal as in (a). Fig. 6.22(c) uses a 'P' channel device requiring a positive control voltage, whereas (b) and (d) use an 'N' channel device which requires a negative control voltage. Zener diodes may be used for level shifting to obtain the correct bias level, or as in (b) and (d) to allow a desired output voltage swing to occur before the control begins to take effect. This allows more flexibility in the achievement of the desired output swing. The speed of response depends on the time constant of the C_1R_1 smoothing components, and the latter determines the ripple of the bias voltage. Capacitor C_2 in circuit (d) has been used to isolate the d.c. feedback from the a.c. feedback in order to achieve the optimum d.c. stability of the amplifier working levels. In circuit (c), R_1 is a potentiometer which can be used for initial level setting. The dependence of output amplitude on ambient temperature is determined by the relationship between FET drain-to-source resistance and gate-to-source voltage.

An interesting possibility is the circuit (e). This uses an integrated-

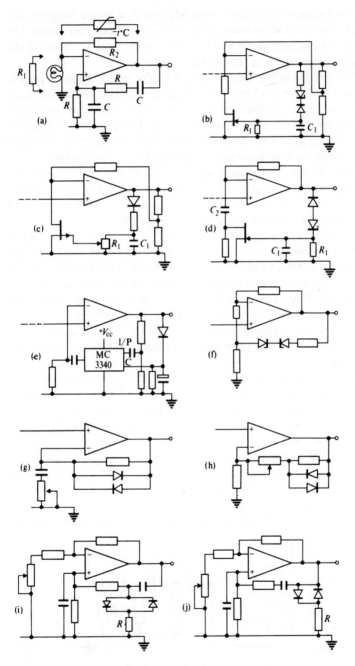

Fig. 6.22 Oscillator AGC Techniques

circuit voltage-controlled resistor MC3340P, a device designed as a remotely operated volume-control circuit for audio applications. The control input terminal is biased at the correct level for inserting the desired attenuation in the feedback path.

The circuits in Fig. 6.22(f)–(h) use diodes in the feedback path to allow a minimum output swing to be generated before the non-linear resistance effects of the diodes begin to increase the feedback signal. As the voltage increases so the resistance decreases, and reduces the closed-loop gain. The abruptness of the diode characteristic changing from high to low resistance would generate considerable output distortion, but this is minimised by connecting resistors in series or parallel with the diodes to smooth the transition. The circuit (h) has been shown to be capable of producing sine waves with less than 0.1 per cent 2nd, 3rd and 5th harmonic distortion.[78]

The last two examples, circuits (i) and (j), again use diodes but this time the negative feedback is not amplitude-dependent, and the control is exerted by the non-linear feedback in the positive loop. In these cases R is used to reduce the abrupt change in conductance of the diodes. It has been shown that the circuit (i) can cover a frequency range of 100 000 to 1 with no more than ± 0.3 dB change of amplitude, by changing the capacitor values.[79] The disadvantage is that changing the potentiometer to set the amplitude also affects the frequency.

It is perhaps worth mentioning here the R–C phase-shift oscillator, which uses a ladder of at least three R–C or C–R cascaded sections. The three sections are required to produce the necessary 180° phase shift so that when the network is connected in the positive feedback loop the condition for oscillation is realised. Either three- or four-section R–C or C–R circuits may be used, but the frequency of oscillation is different for all four combinations. The gain-control feedback is again required as in the above examples. The arrangement is shown in Fig. 6.23. This circuit is not often used, since it uses more components than the Wien bridge and has a higher attenuation for

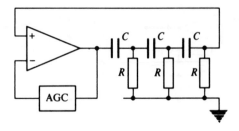

Fig. 6.23 Phase-shift oscillator

180 ° phase shift, thus demanding higher loop gain. The frequency at which 180 ° phase shift occurs can be obtained from Appendix 2, equations [11a], [13a], and the attenuation of that frequency from the expressions for the transfer function $|1/a_{11}|$.

Section type		R–C	C–R	
No of sections	fr	Attenuation	fr	Attenuation
3	$\dfrac{\sqrt{6}}{2\pi CR}$	$\dfrac{1}{29}$	$\dfrac{\sqrt{0.166}}{2\pi CR}$	$\dfrac{1}{29}$
4	$\dfrac{\sqrt{1.43}}{2\pi CR}$	$\dfrac{1}{18.36}$	$\dfrac{\sqrt{0.7}}{2\pi CR}$	$\dfrac{1}{18.36}$

It should be noted that the above discussion on oscillators has assumed that the amplifier itself does not contribute significant phase shift. This may well be true at low frequencies, but as the frequency increases well beyond the amplifier -3 dB point, the amplifier phase shift begins to add to that provided by the R–C network, and will cause the predicted oscillation frequency to differ from that measured. Chart 3.3 shows the phase shift resulting from the compensation of a single-pole roll-off amplifier, and it can be seen that at high frequencies when there is low loop gain available (loop gain $= \beta A$ where A is the amplifier open-loop gain at the frequency of interest, and β would be $\frac{1}{3}$ for a Wien bridge oscillator), the phase shift is very significant. It should also be noted that slew rate will limit the usable oscillation frequency.

A useful reference for those interested in analysing the distortion in the output of phase-shift oscillators is given in reference 80.

6.8.2.2 Twin (bridged)-T oscillators. The bridged *T*-network has been used for many years as a filter network in various forms. It is basically a 'notch' filter having a narrow-band reject-transfer function, and is often used to remove specific unwanted frequency components from signals, e.g., 50 Hz mains. When the network is connected in the feedback path of an amplifier, the resulting circuit will produce a bandpass characteristic. It is only a small step to conceive that a sharp frequency-selective network of this type could be connected in a negative feedback loop, so that this frequency becomes preferred to all others and determines the oscillation when instability has been established. Some alternative networks are shown in Fig. 6.24. The networks of (a) and (b) do not produce a true 'notch' characteristic (i.e. a theoretically infinite attenuation), but produce a strong attenuation

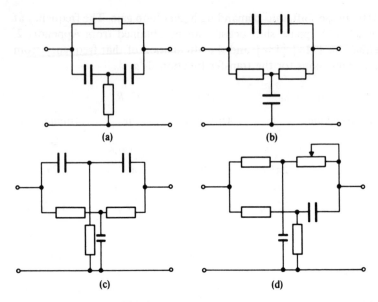

Fig. 6.24 Twin-*T* networks

quite suitable for use as an oscillator. The network (c) is the true notch and (d) shows a method of tuning using a single resistor.[81] The usual analysis for such networks assumes a balanced configuration, a zero source resistance driving-point and an infinite output loading. One method for analysing (c) is given in Appendix 12.

Having connected such a frequency-selective network in the negative feedback path, positive feedback is introduced to establish the oscillation. A useful background to some phase shifting circuits has been given by Butler.[82]

6.8.2.3 Quadrature oscillators. This description is often used to describe a double integrated circuit, connected in a feedback loop so that the output of one integrator produces a sinusoidal output, and the other integrator outputs the same frequency but phase-shifted by $90°$ from the first. A quadrature oscillator is therefore a circuit which can produce sine and cosine outputs simultaneously, but there are several ways of implementing the requirement; Fig. 6.25 shows some possibilities. The circuit (a) is the basic double integrator configuration. Amplifier A_1 is connected as a conventional inverting integrator, whose transfer function is $-j/\omega CR$, thus effectively producing a phase shift of $-270°$ constant with frequency, between points A and B. A_2 is connected as a non-inverting amplifier having the transfer function $(1+j\omega CR)/j\omega CR$ and the R–C network at the non-

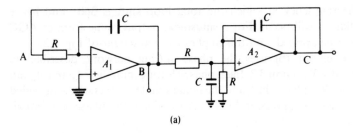

(a)

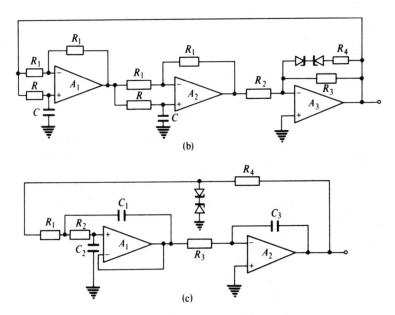

(b)

(c)

Fig. 6.25 Quadrature oscillators

inverting input has the function $1/(1 + j\omega CR)$. The overall result is that the circuit between B and C can produce a maximum phase shift of $180°$. However, only $90°$ is required for $360°$ around the loop, so that the passive network and A_2 are only required to provide $45°$ lag each. For the passive network,

$$\tan \phi = -\omega CR$$

and for A_2,

$$\tan \phi = \frac{-1}{\omega CR}$$

Both of these are satisfied when $\phi = 45°$, i.e. $\tan \phi = 1$. The frequency at which this occurs is therefore $\omega = 1/CR$, i.e. $f = 1/2\pi CR$, and this is the frequency at which the total loop shift is $360°$ and hence oscillation will occur. For a sinusoidal output, some form of AGC should be applied, and this applies to each circuit of Fig. 6.25.

The circuit (b) uses the phase-shifting network described in Appendix 3, section 3.2. Each phase-shifting amplifier A_1 and A_2 can produce $90°$ phase shift, and the remaining $180°$ inversion is provided by A_3. From expression [19] in Appendix 3, the phase-shift circuit gives:

$$\tan \phi = \frac{-2\omega CR}{1 - \omega^2 C^2 R^2}$$

and therefore $\phi = 90°$, when $\tan \phi = \infty$,

$$\text{i.e., } 1 - \omega^2 C^2 R^2 = 0$$

$$\therefore f = \frac{1}{2\pi CR} \qquad \dots [137]$$

It is also shown that the circuit has unity gain irrespective of frequency, so that A_3 requires a gain slightly greater than unity to establish oscillation. This is achieved by making R_3/R_2 slightly greater than unity. When the output swing increases in amplitude so that the zeners conduct, R_4 is switched in parallel with R_3 so that the gain reduces to slightly less than unity. In this way automatic starting is ensured, and the loop gain is maintained at precisely unity to prevent saturation. An advantage of this circuit is that R can be changed to adjust the frequency. The sine and cosine outputs are available at the outputs of A_1 and A_2, but if each resistor R is not equal, this relationship will not apply. In common with all of these oscillator circuits, at higher frequencies the individual phase-shift contributions from each amplifier must be considered, as they will affect the resultant oscillation frequency.

Amplifier A_1 is connected as a conventional Sallen and Key-type two-pole lowpass active filter in circuit (c).[83] Amplifier A_2 is a conventional inverting integrator which provides a total phase shift of $270°$. The circuit therefore requires that A_1 contribute $90°$ phase shift before the circuit will oscillate. An analysis of the filter circuit is given in Appendix 13. From equation [75] in Appendix 13,

$$\tan \phi = -\frac{\omega(C_2 R_2 + C_2 R_1)}{1 - \omega^2 C_1 C_2 R_1 R_2}$$

so that for $90°$ phase shift, $\tan \phi = \infty$ and

$$1 - \omega^2 C_1 C_2 R_1 R_2 = 0$$

$$\therefore f = \frac{1}{2\pi \sqrt{R_1 C_1 R_2 C_2}} \qquad \ldots [138]$$

and this is the frequency at which the circuit will oscillate. If R_4 is significant compared with R_1, the effects on [138] should be considered, although the conduction effects of the zener diodes will produce a non-linear effect on the filter source resistance. The amplitude stability will depend on the temperature coefficient of the clipping diodes, and for best performance, temperature-compensated devices should be used. One advantage of this form of amplitude control is that it responds rapidly to disturbance and does not require the stabilisation time of the capacitor smoothing types such as that shown in Fig. 6.22(b), etc. This eliminates the typical 'bounce' effect observed with slower types of AGC when frequency steps are introduced.

A useful circuit which uses the phase-shifting technique of Fig. 6.25(b) and can be used either as an oscillator or as a multi-function filter has been described by Baril.[84]

6.8.2.4 Crystal oscillators. A quartz crystal will resonate at a frequency dependent on its angle of cut relative to the crystal structure and the geometry. A typical equivalent circuit for a crystal is shown in Fig. 6.26. The circuit has a series resonance (low impedance) and a parallel resonance (high impedance) which can be solved as follows.

Assuming R_x is very small, and C_m is the mounting capacitance,

$$z = \frac{-j/\omega C_m (j\omega L - j/\omega C_x)}{-j/\omega C_m + j\omega_L - j/\omega C_x} \qquad \ldots [139]$$

Solving [139], and equating z to zero,

$$f_1 \approx \frac{1}{2\pi \sqrt{LC_x}} \qquad \ldots [140]$$

and for $z = \infty$,

$$f_2 \approx 1 \bigg/ 2\pi \sqrt{\left[L \frac{C_x C_m}{C_x + C_m} \right]} \qquad \ldots [141]$$

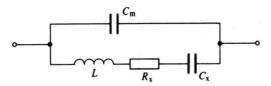

Fig. 6.26 Crystal equivalent circuit

The impedance at f_1 resonance becomes purely resistive, and has value R_x. Therefore if the crystal is used to replace R_2 in the series part of the positive feedback loop of Fig. 6.22(a), and $f_1 = 1/2\pi CR$, the frequency of oscillation will be determined by the very high 'Q' $(\omega L/R_x)$ of the crystal in its series resonant mode. This produces an oscillator with the very high frequency stability of the crystal; the temperature coefficient of frequency can be further improved by operating the crystal in a temperature-controlled oven. Since the crystal impedance varies with temperature, one of the AGC techniques described in section 6.8.2.1 should be utilised.

6.8.2.5 State-variable oscillator. The state-variable circuit is described in more detail in Chapter 8, since its primary function is in active filter circuits. The real advantage of such circuits materialised with the advent of quad amplifiers, since the circuit is based on a multi-amplifier and feedback loop arrangement. A low-distortion sine wave oscillator has been given by Jung,[85] and is shown in Fig. 6.27. The state-variable filter circuit is formed by amplifiers A_1, A_2 and A_3 and the controlled-gain AGC path by A_4. This is a useful arrangement because a three-phase output is available at up to 5 KHz for servo or instrumentation systems. The frequency is given by $f = 1/2\pi CR$, and A_4 inserts a large d.c. gain in the feedback loop which minimises the spread in the operating characteristics of the FET, the latter being

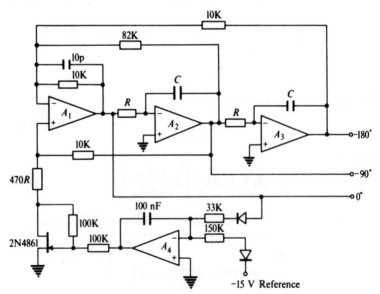

Fig. 6.27 State-variable oscillator

used as a voltage-controlled resistor. The presence of A_4 also has the effect of virtually eliminating the temperature-dependent channel resistance and the change of gate-to-source voltage with temperature of the FET, resulting in good temperature stability of output amplitude.

6.9 Function generators[86]

The accepted definition of a function generator is a circuit which is capable of producing an output voltage which is either: (a) a non-linear function of the input, or (b) a time-varying signal. Another popular method for describing it is a circuit that produces an output signal voltage related to an input signal by some specific or adjustable function. Integrated circuits are available which are designed specifi-cally for use as function generators, and in this context it is usually inferred that triangle, sawtooth and pulse waveforms can be pro-duced. It is often a simple matter to produce pulse waveforms and triangular waves are often available from the same circuits, involving as they do, constant current charging of capacitors. For example, a triangular wave can be obtained from the capacitor in Fig. 6.16(f), from V_1 in Fig. 6.16(g), and V_1 in Fig. 6.20(a). In some cases, buffering is required before external loads are connected so that there is negligible effect on the source, but this is less important where the output is taken from the feedback point since this is normally a low-impedance source.

Sawtooth generators are often useful for process timing and control, and one method of implementation is shown in Fig. 6.28.

Amplifier A_1 is a comparator and A_2 an integrator. If diode D is reverse-biased, A_2 output ramps-up with a slope of $|-V|/CR_4$. When the output reaches the level defined by

$$V_1 = \frac{R_2}{R_1 + R_2} V_o - \frac{R_1}{R_1 + R_2} (V_z + V_{f_z}) \qquad \ldots [142]$$

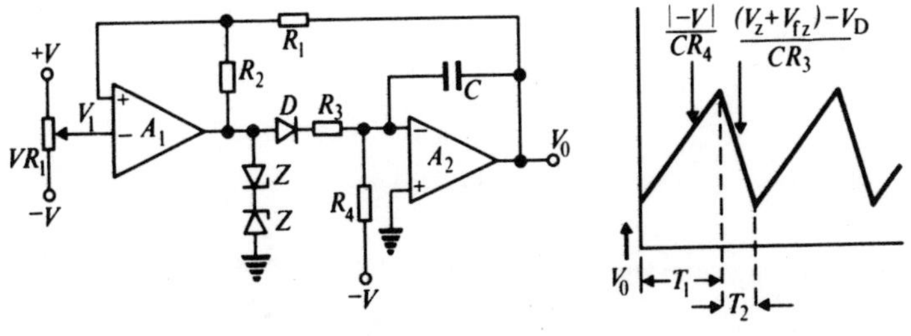

Fig. 6.28 Sawtooth generator

then the comparator output changes to $+(V_z + V_{f_z})$ and diode D is switched into conduction. If $R_3 \ll R_4$, the output ramps down very quickly until V_o is defined by

$$V_1 = \frac{R_2}{R_1 + R_2} V_o + \frac{R_1}{R_1 + R_2} (V_z + V_{f_z}) \qquad \ldots [143]$$

It follows from [142] and [143] that the peak-to-peak value of the output sawtooth is

$$V_{opp} = 2\frac{R_1}{R_2} (V_z + V_{f_z}) \qquad \ldots [144]$$

and the average d.c. level of the output is

$$V_{oav} = V_1\left(1 + \frac{R_1}{R_2}\right) \qquad \ldots [145]$$

The periods can be defined by

$$T_1 = 2\frac{CR_4 R_1 (V_z + V_{f_z})}{R_2 - V} \qquad \ldots [146]$$

$$T_2 = 2\frac{CR_3 R_1}{R_2} \frac{(V_z + V_{f_z})}{(V_z + V_{f_z}) - V_d} \qquad \ldots [147]$$

If $T_2 \ll T_1$, and $-V = V_z + V_{f_z}$, then

$$f = \frac{R_2}{2CR_1 R_4} \qquad \ldots [148]$$

7

Miscellaneous applications

7.1 Transducer amplifiers and signal conditioning

One of the most important applications of linear integrated circuits is to process the output signals generated by various types of transducer. References 87–9 will be found useful for information on the design and use of transducers. Transducers are basically devices for converting one physical variable into another, but in our context are usually taken to imply the conversion of a physical variable into an electrical signal. Sometimes the conversion is reversible (though not necessarily with the same device), for example a pressure transducer will convert pressure to an electrical signal, or alternatively a current can be converted to a pressure for control functions. A wide variety of techniques can be used to convert such variables as flow, temperature, pressure, acceleration, velocity, turbidity, etc. into electrical signals, and may include strain gauges (piezo-resistive devices), piezoelectric crystals, seismic mass, optical, inductive and many other conversion principles. One consideration often underlying the use of transducers is the basic system arrangement, and a little background to this is given in the next section.

7.1.1 Transducer connection methods

Digital data transmission from transducer measurement sites is becoming increasingly important due to the advantages of high accuracy, computer compatibility, ease of multiplexing, and noise rejection. However, analog methods have not been displaced and will continue to be complementary as a result of flexibility, low cost, and reliability in medium-accuracy applications. There is a choice of methods for connecting transducers and the requirements of the system dictate which technique is used. These methods can be conveniently discussed by considering the number of wires which are used to connect the sensor to the associated signal conditioning, either as an integral package or with the two parts separated by long cables.

There is no real limitation to the type of transducer which can be used with these connections; self-generating types such as thermocouples are equally applicable along with externally excited types, for example, thermistors. The differences lie within the circuits used to amplify or process the signals into the required form. In most cases the sensor signals themselves are low-level, and would normally require amplification before they could be usefully applied; the comments here are intended to apply to proportional types of sensor rather than simple on/off devices (for example, limit switches), as the latter are usually capable of producing control signals directly. There are, of course, exceptions such as proximity detectors and serrated-disc magnetic tachometers. The methods will not be discussed in numerical order of wires used, as the concepts will be more apparent by treating the most obvious methods first.

7.1.1.1 Four-wire-connection. This is probably the most common arrangement, especially for externally excited systems. A small proportion of the input power is transmitted to the output terminals and modified in such a way that the output is proportional to the sensed variable. Figure 7.1(a) shows the connection. Typical realisations are shown in Fig. (b) and (c). The input and output may be any combination of a.c. or d.c. voltages or currents, and the associated circuit design will reflect this. Examples of (b) would be strain gauges or thermistors, and (c) would be variable-reluctance or differential-transformer transducers.

Sometimes it is desirable to avoid ground-loop problems, in which case the input terminals are isolated from the output terminals in the associated circuits. When the sensor must be located some distance from the electronics, the lead resistance can cause difficulty, and is the usual reason for use of the six-wire connection described next.

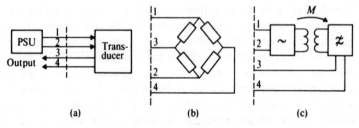

Fig. 7.1 Four-wire transducer connection

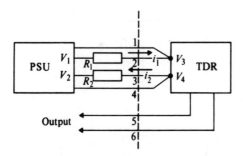

Fig. 7.2 Six-wire transducer connection

7.1.1.2 Six-wire connection. Refer to Fig. 7.2. The transducer excitation current flows from the source through the lead resistance R_1 and produces V_3 at the transducer. Likewise the return current produces a potential drop across R_2, and so the transducer excitation voltage depends on lead length. Resistors R_1 and R_2 are also subject to the typical temperature dependence of copper wire of 0.4 per cent/°C.

One method of eliminating the difficulty is to excite the transducer from a constant current source; however, this can cause difficulty if the transducer resistance varies with temperature because the input voltage will then vary, and in arrangements where the output is directly dependent on input voltage an unwanted temperature coefficient results. Constant-current four-wire circuits are often used in data logging systems because of the ease of implementation and compatibility with scanning techniques. Some transducers (for example semiconductor strain gauges) can be arranged to exhibit integral temperature compensation by utilising the change of terminal voltage when excited from a constant current.

Voltages V_3 and V_4 are monitored by leads 1 and 4, and assuming that each wire is terminated by a high input resistance at the PSU, negligible current flows in the wires and the voltages at the PSU are V_3 and V_4. Then $V_1 - V_3$ and $V_2 - V_4$ are used as error signals such that V_3 and V_4 are driven to the required levels. In this way the transducer excitation is maintained at a constant voltage irrespective of the values of R_1 and R_2.

7.1.1.3 Three-wire connection. The most obvious connection is the potentiometric transducer, which is used for a wide range of applications. Typical examples would be displacement, pressure, angular position, and level-detection measurements. Figure 7.3 shows the connection, and also the method by which a four-wire type of

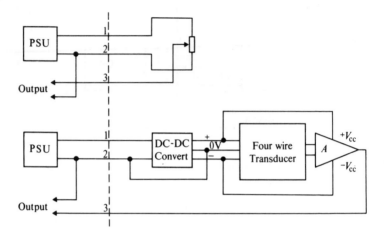

Fig. 7.3 Three-wire transducer connection

transducer can be connected into a three-wire system. This is a popular arrangement where a different type of transducer or measurement is required on an existing experiment. The d.c.–d.c. converter produces a negative supply voltage with respect to the common, so that the amplifier output terminal can go down to the common potential, thus simulating the potentiometric diagram above. There are single-supply integrated-circuit amplifiers available where the outputs can go to the common line, but the actual output voltage depends on the current sink due to the active nature of the pull-down output (see Chapter 9).

7.1.1.4 Two-wire connection. Probably the most common arrangement in industrial control systems, two-wire transducer connections have the following advantages:

1. Accurate data transmission using current rather than voltage.
2. Low installation costs; minimum cable requirement and low current operation.
3. Low power consumption.
4. Reduction of the probability of incendive faults in hazardous areas; ideally suited to intrinsic safety applications.[90–3]
5. 'Live' zero provides rapid fault situation indication.

British Standard BS3586:1970 'Analogue direct current signals for telemetry and control', specifies the preferred loop current as 4 to 20 mA. Figure 7.4 shows the arrangement. The initial 4 mA is used to excite the transducer and power any necessary circuits, and the signal is superimposed so that the full-scale transducer output is 20 mA and

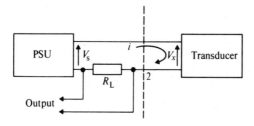

Fig. 7.4 Two-wire transducer connection

the span is 16 mA. This connection clearly requires a low power-consumption design for the transducer and electronics. Although this method has several advantages, the principal disadvantage is the necessity to cope with the 'live zero' at the receiving end of the line; however, techniques for overcoming this problem have been discussed elsewhere.[94]

7.1.1.5   One-wire connection. This is, of course, a misnomer; it simply refers to the fact that one wire of a pair is replaced by an earth return. This is commonly used with field telephone systems, but remains an interesting possibility for transducer data transmission using some form of current or frequency modulation. The main difficulty is likely to be the high-impedance nature of the return which thus demands a high-voltage power supply. Recent improvements in two-wire current-transmission circuits indicate the possibility of using this technique in the future; its simplicity and low installation cost will be a spur to progress.

The only possibility that has not been mentioned is a five-wire arrangement: this does not appear to offer any particular advantages over the other connections, but may be useful on occasions. The system designer thus has many alternatives open to him; the decision about which connection to use will be determined by such considerations as accuracy, complexity and economics.

7.1.2 Strain-gauge amplifiers

Many of the drift-corrected amplifier techniques described in section 5.13 are used for this purpose. In some cases where highest accuracy is paramount but bandwidth can be restricted, for example, load cell and weighing applications, the auto-zero technique is popular. In others, where rapid response is necessary, for example in the dynamic measurement of pressure transients, optimised linear d.c. amplifiers are utilised. The choice of amplifier therefore depends on the

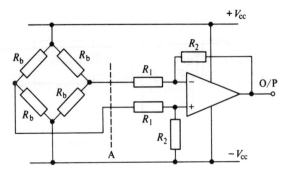

Fig. 7.5 Simple bridge output amplifier

application rather than on the type of signal source. There are, however, some generalised rules which can be drawn up for strain-gauge amplifiers.

Strain gauges usually represent low-impedance signal sources, i.e., voltage generators (apart, perhaps, for some rarely used semiconductor gauges), so that we are usually more concerned with offset voltage and drift, and voltage noise generation rather than their current equivalents. However, for the ultimate in performance, the effect of offset current in the source must be considered, especially when extra resistors are connected in series with the input terminals. The simplest bridge amplifier could be represented as in Fig. 7.5, and the gain would be

$$\frac{R_2}{R_1 + R_b/2}$$

An obvious question is: if the addition of R_1 increases the source resistance and therefore the current-generated offset, why not remove R_1 and let the source resistance determine the gain? The answer is that the gain would be dependent on variations of R_b, and since R_b changes with temperature, so also does the gain. In fact, semiconductor strain gauges often have very large temperature coefficients of resistance and the resulting gain instability would be disastrous. In addition, the differential input resistance at A in the circuit is $2R_1$. If both R_1's are removed, then the approximate input resistance is $\dfrac{2R_2}{A}$ where A is the open loop gain. It is therefore important to consider the input loading effect on the bridge output.

Another most important consideration is that of common-mode signals. The bridge CM output must, of course, be within the CM

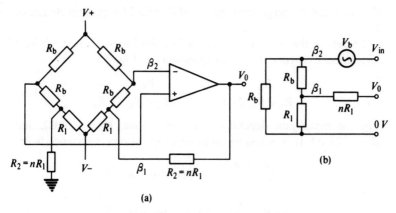

Fig. 7.6 Low source resistance bridge amplifier

range of amplifier (depending on the ratio of R_2/R_1), and common-mode rejection must be optimised so that changes of bridge CM output level with temperature do not cause significant differential errors. This problem is discussed in Chapter 4.

One way to avoid the penalty of increased current-generated drift from R_1 without making the gain totally dependent on R_b is shown in Fig. 7.6.

Referring to Fig. 7.6 (b),

$$\beta_1 = \frac{1}{1 + n(1 + R_1/2R_b)} \qquad \ldots [149]$$

$$\beta_2 = \tfrac{1}{2} \text{ and gain} \approx \frac{1}{\beta_1 \beta_2}$$

$$\therefore G = 2 + n\left(2 + \frac{R_1}{R_b}\right) \qquad \ldots [150]$$

Variation of G with R_b,

$$\frac{\partial G}{\partial R_b} = -\frac{nR_1}{R_b{}^2}$$

and percentage change of $G = \dfrac{\partial G}{G}100$

$$= -\frac{nR_1}{R_b[2 + n(2 + R_1/R_b)]} \times \text{percent change of } R_b \qquad \ldots [151]$$

For example, if $n = 100$, $R_1 = 100\,\Omega$ and $R_b = 350\,\Omega$, $R_2 = 10\,k\Omega$, then $G = 230.6$, and percent change of $G = 0.124 \times$ percent change of R_b.

Therefore the overall gain has a low dependence on the bridge resistance. Compare the gain expression for Fig. 7.5,

$$G = \frac{R_2}{R_1 + R_b/2}, \frac{\partial G}{G} = -\frac{R_b}{2(R_1 + R_b/2)} \frac{\partial R_b}{R_b} \quad \ldots [152]$$

For the same gain dependence as that above, $R_1 = 1.24 \, k\Omega$, and $R_2 = 326 \, k\Omega$. It is necessary to increase the source resistance from

$$\frac{R_b(R_b + R_1)}{2R_b + R_1} = 197 \, \Omega \quad \text{for Fig. 7.6}$$

$$\text{to } R_1 + \frac{R_b}{2} = 1.4 \, k\Omega \quad \text{for Fig. 7.5}$$

This means that with the circuit of Fig. 7.6 it is possible to keep the source resistance very low, produce high gain, and eliminate the need for high-value feedback resistors which can be inconvenient. If a fixed-bridge excitation voltage is used, the reduction of bridge sensitivity from

$$\frac{\delta R_b}{R_b} \times V_b \text{ for a four-active arm bridge}$$

(i.e. gauge factor × strain × bridge voltage),

$$\text{to } \frac{\delta R_b}{R_b + R_1/2} \times V_b \text{ for Fig. 7.6 must be considered.}$$

The above examples illustrate the care that is required when designing even the simplest high-performance amplifier. The source-resistance problem is largely overcome by the use of the instrumentation amplifier (section 7.1.3) where the gain can be determined by a single resistor independently from the source. There are, of course, many possible configurations in which several op amps can be used, and some of these are discussed in reference.[95] One popular arrangement is shown in Fig. 7.7.

$$V_1 = \frac{R_1 + R_2}{R_1} \frac{V}{2}, \; V_o = -\frac{R_4}{R_3} V_1 - \frac{R_3 + R_4}{R_3} \frac{V}{2}$$

$$\therefore V_o = -\frac{R_4}{R_3} \frac{R_1 + R_2}{R_1} \frac{V}{2} - \frac{R_3 + R_4}{R_3} \frac{V}{2}$$

$$\text{Gain} = \frac{V_o}{V_{in}} = \frac{V_o}{V} = -\frac{1}{2}\left[\left(1 + \frac{R_4}{R_3}\right) + \frac{R_4}{R_3}\left(1 + \frac{R_2}{R_1}\right)\right] \quad \ldots [153]$$

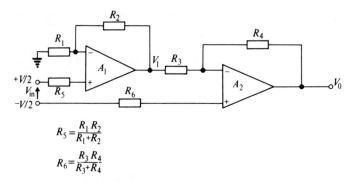

$$R_5 = \frac{R_1 R_2}{R_1 + R_2}$$

$$R_6 = \frac{R_3 R_4}{R_3 + R_4}$$

Fig. 7.7 Differential amplifier gain independent of source resistance

If $\dfrac{R_2}{R_1} = \dfrac{R_4}{R_3} = K$,

$$G = \frac{K^2 + 2K + 1}{2} \qquad \ldots [154]$$

or if $\dfrac{R_1}{R_2} = \dfrac{R_4}{R_3} = N$,

$$G = 1 + N \qquad \ldots [155]$$

To optimise CMRR it is mandatory to use expression [155]. The above arrangement provides high input resistance and independence of gain from the source resistance. A potential disadvantage is that the full common-mode input range operates on both input terminals of each amplifier and the circuit is therefore liable to CMRR-based errors (apart from the external resistor tolerance errors); however, dual amplifiers are available offering matched CMRR and offset voltage drift (e.g. OP–04), and these can provide very good performance in this circuit.

Strain-gauge bridges often require external balancing owing to initial resistance tolerances, imbalance of residual strain, mismatch of creep with time between each arm, stress relief in the substrate or the desire for zero suppression facilities. It is often adequate to provide direct control at the bridge as in Fig. 7.8.

Points to consider are the loading effects of the balance components on the bridge output, the temperature coefficient of resistance of these components, the range of bridge common-mode output voltage and the change of common-mode voltage with temperature (circuits (a) and (b)). The circuit (c) demands the use of a very low-resistance potentiometer (and hence low resolution when wire-wound

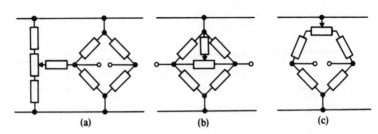

(a)   (b)   (c)

Fig. 7.8 Bridge balancing

potentiometers are used for their low temperature dependence) or slide wire, and contact resistances can cause instability or thermoelectric voltage errors. Some of the problems experienced with strain gauges have been described in reference 96. It is usually better to use fixed resistors rather than potentiometers if possible, but in circuit (c) the temperature coefficients of resistance of the resistors on each side of the bridge should ideally be very low and track each other.

Whilst the above balancing procedure is often adequate for wire or metal foil/film gauges (and in many cases the balance process can be used to compensate for initial amplifier offset error simultaneously), the increasing popularity of semiconductor gauges presents some difficulty. Diffused semiconductor gauge bridge arms are often very well matched for TCR (temperature coefficient of resistance) and therefore the bridge output is stable with temperature, but the change of bridge resistance with temperature makes it difficult to use direct-balancing circuits. In many cases, the op amp external null-balance control can be used to offset bridge imbalance, but the effects on amplifier drift should be considered (see section 4.2.2). Often a better alternative is to use the circuit arrangement shown in Fig. 7.9. The preamplifier gain is selected so that it can handle the expected dynamic range of the input signals, and balanced tracking resistors are used for best performance. The zero balancing is then applied to A_3, and has the advantages of operating at a high signal level and not interfering with the bridge source or the preamp input offset. Some compromise may be required to optimise CMRR, unless a scheme such as that shown in Fig. 4.16 is used. Dual-tracking voltage regulators can be used for power supplies in order to optimise the setting stability, CMRR and power supply rejection ratio.

Gain control in Fig. 7.9 is shown as a variable attenuator. This is often a desirable method since it allows changes of gain to be obtained after balancing to 0 V at the output of A_3 without affecting the zero setting (a common failing of simple op amp circuits). It is also easier to make the resistor op amp circuits). It is also easier to make the resistor

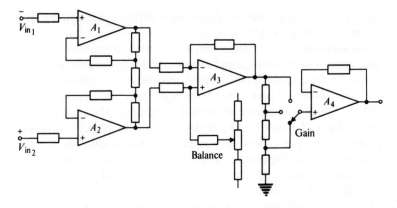

Fig. 7.9 Resistance bridge transducer amplifier

chain a low resistance source so that changes in the offset of A_4 due to offset current in the source can be minimised.

A very high-performance amplifier can be built by the method shown in Fig. 7.10. This utilises the very low drift and low noise of the multi-device matched pair of transistors used as the first stage of a high-gain amplifier. At the time of writing this matched pair of transistors can offer a better front-end specification than any available

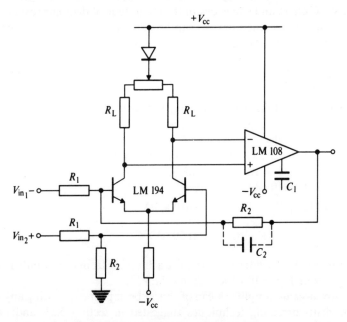

Fig. 7.10 Low-noise, low-drift amplifier

op amp input stage, but no doubt such performance will become available in integrated form in time.

An important consideration with this circuit is to estimate the input drift of the op amp (voltage and offset current in the source resistance R_L) and refer it to the input terminals divided by the voltage gain of the transistors, where the first-stage gain is approximately $10^3 R_L I_e/26$. It is therefore important to minimise the op amp drift and noise and maximise the first-stage gain to obtain optimised performance. Total noise increases with bandwidth, so if wide bandwidth is not required the amplifier can be followed by a filter (or the amplifier bandwidth itself restricted). The overall gain is conventionally R_2/R_1, and all the precautions concerning resistor noise, temperature coefficient, tracking and tolerance must be observed.

The input offset current, which is determined by the bias current required to support the chosen collector current and the device matching, flows in R_1 and produces offset voltage and drift. The frequency response of the transistors will add an extra high-frequency pole to the open-loop response, and the compensation capacitor C_1 must be chosen with this in mind to obtain overall stability; C_2 may be used to insert phase advance into the loop and produce a greater stability margin. All thermoelectric and thermal-gradient sources must be minimised; a typical solder-to-copper junction can generate $3 \mu V/°C$ error, and some idea of the magnitude of drift improvement provided by matched devices can be obtained from the following.

For a bipolar transistor,

$$V_{be} = \frac{KT}{q} \log_e \frac{I_c}{I_s} \qquad (I_c \approx I_e) \qquad \ldots [156]$$

where K = Boltzman's constant, 1.38×10^{-23} J/K
         T = absolute temperature, $298\,°K$
         q = electron charge 1.6×10^{-19} coulomb
         I_s = base emitter diode saturation current, typically
             5×10^{-14} A at $25\,°C$
         $I_c = 10\,\mu A$

From [156] $V_{be} \approx 0.49\ V = 490\,000\ \mu V$

and $\left(\dfrac{\partial V_{be}}{\partial T}\right)_{I_s} = Kq \log_e \dfrac{I_c}{I_s} \approx 1650\ \mu V/°C$

Compare this with the performance achievable from the amplifiers of Fig. 7.9 or Fig. 7.10 of less than $0.5\,\mu V/°C$.

Strain-gauge amplifiers, can of course be implemented using any of the drift-correcting techniques suggested in section 5.13, and one possible arrangement is outlined in Fig. 7.11. In principle it is an auto-

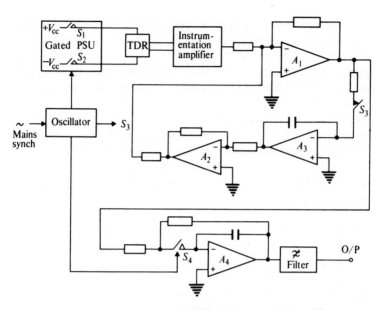

Fig. 7.11 Drift-corrected transducer amplifier

zero technique with the correction signal applied after the pre-amplifier. The gated power supply provides the bridge excitation, and the oscillator drives the switches with the required synchronisation. Assume initially that S_1 and S_2 are open, and S_3 is closed. The integrator A_3 ramps at a rate dependent on the total offset error referred to the input terminals. The advantage of gating the bridge supply rather than the signal source is that the source resistance is always connected, so any offset current generated errors will be included in the correction.

S_3 remains closed for a short time, just long enough for the ramp to reach the error voltage. S_3 then opens, and the output of A_1 is held at 0 V by the voltage held on the integrator. S_1 and S_2 close and the transducer produces the signal output, and after a suitable settling time S_4 closes and the leaky integrator A_4 follows the signal. The next cycle starts when S_4 opens followed by S_1 and S_2, and the signal is held by A_4 until S_4 closes again after the next auto-zero sample. The output filter removes switching transients from the signal.

A scheme like this is often synchronised to mains frequency so that integration takes place over an exact number of cycles and the average is zero; high levels of mains pick-up can be rejected in this way. The disadvantage is that bandwidth is very limited, but this is often not a problem with many strain-gauge applications, e.g., load-cell weighing, pressure transducers, structural stress analysis, etc.

Excitation to strain gauges is often a.c. in form, and the amplifier can then be a.c.-coupled, thus eliminating the d.c. drift problem. The a.c. waveform, amplitude-modulated by the signal, is detected by a phase-sensitive detector.[97-8] Very high stability can be obtained in this way,[99-100] but the circuits tend to be more complex because of the need for quadrature balancing and the necessity to minimise amplifier phase shift. In addition, the transducer a.c. excitation must be accurately amplitude-stabilised, a requirement which is often more difficult to implement than with d.c. excitation. Carrier systems are, however, in extensive use and offer an alternative to the d.c. methods described above. The carrier system minimises thermoelectric errors, and can provide good dynamic response provided the carrier frequency is high enough. Care is required with long cables to balance the core-to-core capacity so that the imbalance of reactive current is within the range of the amplifier quadrature balance control. The ease of transformer coupling in carrier systems is very useful in applications where signal isolation is required to avoid earth loop currents, and where the signal source is at a high voltage compared with the monitoring circuit (see also section 7.1.4).

7.1.3 Instrumentation amplifier

Although the arrangement of op amps in Fig. 7.9 is often classified as an instrumentation amplifier, it really describes a circuit function rather than its method of implementation. One manufacturer defines the term as follows: 'an instrumentation amplifier is a *closed loop* gain block with high differential input impedance and an accurately predictable gain relationship between the input and output. By contrast an op amp is an *open loop* gain block which is used to power an external feedback network to achieve the desired relationship between the circuit output and its input.' The circuit of Fig. 7.9 is therefore one way to achieve the objective using separate amplifiers, often in the same package. Another alternative is the totally integrated amplifier principle as exampled by the AD520: Fig. 7.12 shows the operation. I_1–I_4 are voltage-controlled current sources; the action is such that A_2 forces a current dependent on $V_{sense} - V_{Ref}$ determined by R_{scale} to be proportional to the differential input voltage $V_1 - V_2$ acting across R_{gain}. This leads to the expression

$$\frac{V_{sense} - V_{Ref}}{V_1 - V_2} = \frac{R_{scale}}{R_{gain}} = G \qquad \ldots [157]$$

When the sense terminal is connected to the output and the reference

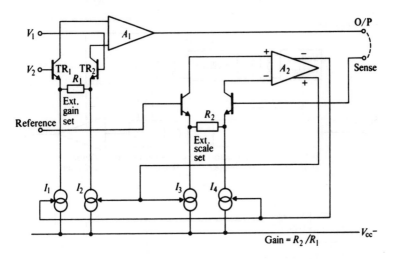

Fig. 7.12 Instrumentation amplifier (AD520)

to 0 V, [157] becomes

$$V_{out} = \frac{R_{scale}}{R_{gain}} (V_1 - V_2) \qquad \dots [158]$$

The device uses thin-film resistors deposited on the surface of the active silicon chip. These resistors offer a 10:1 improvement in temperature-coefficient tracking compared with diffused silicon resistors, and also conserve chip area since their 'sheet' resistance in $\Omega/\square$ is ten times greater than the diffused components.

The reference terminal can be used to advantage in many applications for biasing the output terminal at a desired voltage independently of the gain, and external current bias boosting can be incorporated by connecting the sense terminal outside the boost circuit. For a more detailed description of the operation, see reference 101.

7.1.4 Isolation amplifier

As pointed out in section 7.1.2, carrier amplifiers are convenient for providing input-to-output signal isolation – and this of course means that the output power-supply circuit must be isolated from the input power supply. In other amplifiers isolation can be provided by transformers or opto-isolators.[102] In the case of the latter, solid-state optoelectronic devices usually have large temperature coefficients of their transfer characteristics, and in some circuits pairs of matched

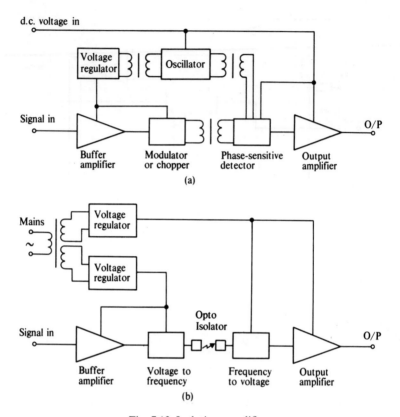

Fig. 7.13 Isolation amplifiers

devices have been used to provide compensation. However, it is usually preferable to use a signal-modulation process such as voltage-to-frequency conversion or analog-to-digital (not amplitude modulation) so that the signal transfer is independent of the slope of the device transfer characteristic. The block diagrams of two possible alternatives are shown in Fig. 7.13. In circuit (a), transformers are used to provide the isolation, and the signal information is translated to an amplitude-modulated carrier; the use of a phase-sensitive detector renders the synchronous demodulation effective against unwanted noise components and prevents reactive errors from causing non-linearity.

Circuit (b) uses an opto-isolator, many of which are available in packages capable of withstanding 5 kV between emitter and detector, and most of which have been developed for digital data transmission. When isolating high voltages, the power supply transformer and the physical construction must also withstand the necessary peak voltages. The key to accuracy with such a system is the accuracy of the V–f and f–V conversion circuits, and particular attention must be paid to non-linearity and drift. Some V–f converters have a speed of response which depends on the signal amplitude, especially where 0 V input produces zero output frequency. The rise time or bandwidth to low-level signals is therefore slow, but this problem can be remedied by producing a positive d.c. offset in the input circuit, thus superimposing the signal on a d.c. pedestal and producing an initial output starting frequency of conversion.

7.1.5 Three-wire circuits

The usual arrangement is discussed in section 7.1.1.3. Op amps specially designed for single-supply operation can be used for these applications. It is important to note that the input terminals can be used at 0 V, but that the output terminal is usually near to, but not at 0 V. The output voltage also varies with current sourced or sinked, so that it is not possible with a single supply to adjust the output through zero, a requirement which is often essential for transducer amplifiers. An output which starts at 0 V may require the capability of moving below zero when a superimposed transducer temperature-dependent zero shift is superimposed, otherwise an initial deadband will result when the transducer is exercised. This requirement inevitably leads to the necessity of providing an 'internal' d.c.–d.c. converter to generate a negative bias voltage. Two simple circuits for generating a negative voltage from a positive without using coils have been given in references 103 and 104, and the circuit of Fig. 7.14 shows a possible arrangement using a multivibrator. The output from the multivibrator is a zero-based positive-going square wave; C_1, D_1 clamp the waveform so that it is negative-going and D_2, C_2 rectify and smooth to

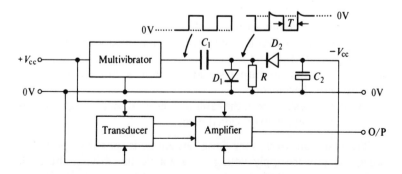

Fig. 7.14 Generating a negative reference

produce a negative reference voltage. It is only necessary for the output part of the amplifier to have a negative capability, and if the voltage is small a low value of current sink is required (depending on the minimum output load). R prevents d.c. charging of C_1. C_1 needs to be of sufficient capacity to prevent the rail current from causing too much droop in the waveform:

$$\left(\% \text{ droop} = 100 \times \frac{T}{C_1 R_{\text{eff}}}, \text{ where } R_{\text{eff}} = -\frac{V_{cc}}{i_{cc}} \right)$$

and the value of C_2 will determine the output ripple:

$$V_{pp} \approx \frac{T i_{cc}}{C_2}$$

7.1.6 Two-wire circuits

The advantages of two-wire circuits have been outlined in section 7.1.1.4. The 'live' zero current of 4 mA must be adequate for supplying transducer excitation (in the case of non-self-generating types) and power to all of the active devices in the circuit, including the ability to preset 'zero' at 4 mA. It is clearly obligatory to examine carefully device tolerances if repeatable circuits are to be constructed. It is apparent that only a very small current will be available for transducer excitation, and this often means that low signals will result which demand low-drift amplifier design. Micropower IC amplifiers are very useful for this type of application. Points of fundamental importance which need initial consideration include:

1. Proportions of the 4mA to be allocated to transducer and electronics.
2. Minimum and maximum voltages across the circuit input.
3. Load driving capability (the higher the line resistance, the higher the supply voltage required. Minimum supply $= V_x + 20\,\text{mA}\ R_L$, where V_x = circuit terminal voltage, R_L = line + load resistance).
4. Dependence of circuit input current on applied voltage.
5. Signal level, drift and noise.
6. Whether input-to-output signal isolation is required (often the case with thermocouples).
7. Insulation resistance of the cabling to be used; leakage between the two cores represents a signal error.

Transmission over very long distances is possible providing adequate power-supply voltage is available for driving the cable resistance. Figure 7.15 shows some system diagrams of circuits that

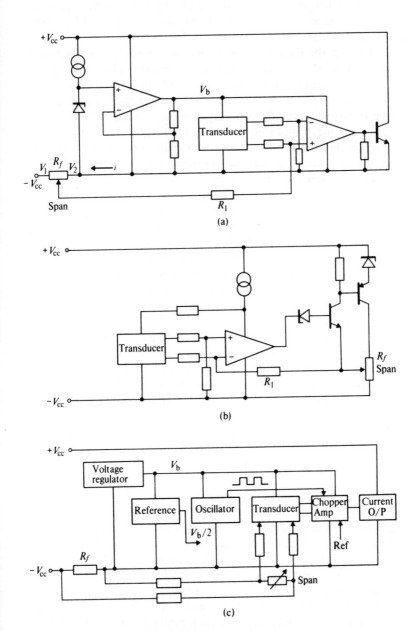

Fig. 7.15 Two-wire circuits

have been used in two-wire transmitters. There are some interesting points to note in these arrangements which directly affect their accuracy and the ease of application.

In circuit (b), the transducer excitation current and the amplifier supply current flow separately into the loop. They add to the controlled output current flowing in R_f to form the initial 4mA. This means that if either transducer current or amplifier current change due to temperature effects or input voltage changes, a zero error results. Some transducer and circuit packages are temperature-compensated as integral units during manufacture, which means that these errors can be minimised. Adjusting the span control has the effect of changing the zero reference as well, since the d.c. conditions on each side of R_1 are not identical.

The circuit of 7.15 (a) is a much better solution, since the whole of the circuit current flows through R_f and the feedback voltage developed at the wiper of the span control ensures that any current variations are controlled within the feedback loop. A hybrid integrated circuit with this configuration is available;* the advantages of building an op amp-based circuit include flexibility of bridge voltage and supply current, choice of low drift-signal amplifiers and the use of precision resistors.

The loop current i produces a voltage drop across R so that V_1 is less than V_2. The feedback voltage is therefore at a negative d.c. level compared with the circuit common line, which means that when the initial condition ($i = 4\,\text{mA}$, usually set by adjusting the bridge output) applies there is a d.c. voltage across R_1. If the span control is now adjusted, this d.c. condition changes, so that the 4mA also changes. This interaction necessitates the cycling of the transducer between zero and full scale whilst adjusting the null and span controls to obtain the 4–20 mA span. When the output current is 20 mA and the terminal voltage is, say, 20 V, the circuit is dissipating 400 mW. Most of this will be at the output stage (over 16 mA flows in the output stage), so the circuit arrangement must take into account the possibility of thermal feedback producing drift in the signal amplifier. This is especially relevant with the hybrid circuit form. Adjusting the 'zero' 4mA output by nulling the bridge does not affect the span; so once the full-range output has been set, the zero reference can be changed without the need for recalibration.

The circuit of Fig. 7.15(c) also encloses all of the circuit within the feedback path to establish a high-accuracy loop current. A low-current multivibrator is used to drive the chopper amplifier to process

---

* LH0045 National Semiconductor Corp.

the low-level transducer output signals. The output current-drive stage is connected to provide the required span current with the correct sense for the negative feedback voltage generated across R_f. The feedback signal is applied to a transducer bridge connected in the manner shown in Fig 7.6(a), and is balanced in nature so that the transducer loading is also balanced in order to avoid common-mode problems. This circuit suffers from span/zero interaction for the same reasons as described above.

A solution to some of the difficulties described above is to be found in the circuit of Fig. 7.16. A_1 amplifies the bridge output, and initial zero errors are balanced at A_2 using R_1. The output of A_2 is at the reference voltage $V_b/2$ after the initial balancing procedure so that both ends of R_2 are at identical d.c. levels. The span control R_2, which is a variable attenuator (see Fig. 7.9) can then be adjusted without changing the circuit d.c. conditions. The feedback voltage is developed by the loop current flowing in R_f. When the initial 4 mA is flowing, V_1 is less than V_2, but the inverting input of A_3 can be adjusted to the same d.c. level as the non-inverting input by the preset R_3. This attenuates the overall feedback voltage, but the loop gain of the output stage is still very high so that accuracy is maintained; when the 4 mA has been set by R_3, the differential input voltage at A_3 is very small, and the span control can now be adjusted without affecting the zero reference. This circuit arrangement means that setting the 4 mA initial current is a once-for-all process;

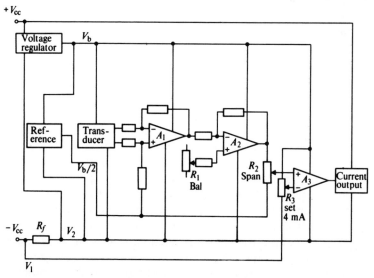

Fig. 7.16 Improved two-wire transducer amplifier

transducer input errors are balanced by R_1 after the presetting operation.

All the above circuits demand the use of low-current amplifiers, since current conserved in the active circuits becomes available for exciting the transducer. More current in the transducer means a higher output signal and thus minimises the amplifier drift limitations. In the circuits of Fig. 7.15 (a) and (b), the bridge nature of the signal source is relied upon to produce the necessary amplifier common-mode input voltage. This makes them difficult to use for single-sided inputs. However, the circuits of Figs. 7.15(c) and 7.16 have a split-rail reference, to which single-sided inputs can be connected. The voltage regulator determines the stability of the transducer output, and low current, low temperature coefficient devices should be used. Such regulators are not difficult to design, and will operate over a very wide input voltage range.[105 – 6] An application of a typical two-wire circuit to an intrinsically safe transmitter for process control applications has been described elsewhere by the author.[107] Other two-wire circuits are described in the application note on the LM10 amplifier.[108]

The above discussion relates specifically to d.c. current transmission. It is also possible to supply d.c. current over two wires to energise the circuits, and to transmit the information back along the two wires superimposed on some form of a.c. carrier. The a.c. component is then separated from the d.c. at the receiving end of the line. These techniques are further discussed in reference 75.

7.1.7 Charge amplifiers

The piezoelectric effect of various crystalline materials is used for many different types of transducer. We may include dynamic pressure transducers, accelerometers, ultrasonic devices, microphones and record styli, and force sensors. Basically the sensors are capacitors, which produce a surface charge as a result of mechanical deformation, and the process is reversible. The signal source is therefore a capacitor and at low frequencies represents a very high impedance. If such a transducer is connected to a voltage amplifier, the input resistance of the amplifier causes charge to leak away from the transducer and thus limits its low-frequency response capability. The operation of piezoelectric transducer is often described in detail in manufacturer's literature, to which the reader is referred, (e.g., 109). A practical arrangement of a transducer connected with a cable to an amplifier input might be represented as in Fig.7.17(a). C_T is the transducer capacitance, C_c the cable capacitance, and C_{in}, R represent the

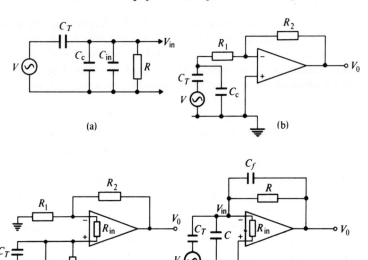

Fig. 7.17 Piezoelectric transducer amplifiers

amplifier input. It is obvious that the load capacitors attenuate the transducer output and R limits the low-frequency response. The bandwidth is derived as follows:

$$\frac{V_{in}}{V} = \frac{R}{R(1 + C/C_T) - j/\omega C_T} \qquad \text{where } C = C_c + C_{in}$$

$$\left|\frac{V_{in}}{V}\right| = \frac{R}{\sqrt{R^2(1 + C/C_T)^2 + 1/\omega^2 C_T^2}} \qquad \dots [159]$$

At the $-3\,\text{dB}$ bandwidth,

$$\left|\frac{V_{in}}{V}\right| = \frac{1}{\sqrt{2}}$$

$\therefore$ [159] becomes

$$f = \frac{1}{2\pi C_T R\sqrt{2 - (1 + C/C_T)^2}} \qquad \dots [160]$$

If $C = 0$ we have the basic expression

$$f = \frac{1}{2\pi C_T R}$$

It is clear that to minimise f (i.e., extend the response to the lowest possible frequency), C must be as small as possible and R very large. The circuit of Fig. 7.17(b), then, is unlikely to be very useful since the input resitance R_1 is limited. The voltage follower (Fig. 7.17(c)) is much more useful, and can be conveniently implemented using FET input amplifiers. The resistor R_b must be inserted to provide a return for the input bias (leakage) current, and it is important to remember that the bias current often doubles for every $10\,^\circ C$ rise in temperature. If no bias return is provided, it will simply charge the capacitance and produce a continuously changing drift at the output. The input resistance R_{in} is very high if the correct devices are chosen, and is further enhanced by the effect of the negative feedback, so that in practice R_b is the effective value, providing precautions are taken with the circuit layout to minimise leakage. Some useful advice is contained within reference 110, including a method for biasing the input terminals so that 'zero' bias current is seen by the source. The amplifier input capacity, although usually very small, may also affect the performance. The transducer is usually specified in terms of its charge per unit of the applied parameter, e.g., force or acceleration, so that for the equivalent circuit of (a), $S = q/g$, where q is the charge produced per unit g. Now, since $q = C_T V$, the measured quantity is $Q = VC_T/S$. From [159],

$$|V_o| = |GV_{in}| = \frac{GR\,|V|}{\sqrt{R^2(1 + C/C_T)^2 + 1/\omega^2 C_T^2}} \qquad \ldots [161]$$

and G for the circuit (c) is $(R_1 + R_2)/R_1$.

$$\therefore Q = \frac{C_T V_o}{SGR}\sqrt{R^2\left(1 + \frac{C}{C_T}\right)^2 + \frac{1}{\omega^2 C_T^2}} \qquad \ldots [162]$$

$$= \frac{V_o}{GS}\sqrt{R^2(C + C_T)^2 + 1/\omega^2}\,/R \qquad \ldots [163]$$

If R is very large,

$$Q = \frac{V_o(C + C_T)}{GS} \qquad \ldots [164]$$

However, it is apparent that the amplifier input signal V_{in} is dependent on the cable capacity, since if R is very large, from [159],

$$|V_{in}| = \frac{V}{1 + (C_c + C_{in})/C_T} \qquad \ldots [165]$$

The cable capacity not only affects the bandwidth but also attenuates the signal.

Consider now the circuit in Fig. 7.17(d)

$$V_x = \frac{RV}{R(1 + C_x/C_T) - j/\omega C_T} \qquad \ldots [166]$$

Where V_x is the effective input voltage (see Section 1.8.3), and $C_x = C + C_f$ and $C = C_c + C_{in}$

And

$$\frac{1}{\beta} = \frac{1 + j\omega R(C_f + C_y)}{1 + j\omega C_f R}, \quad \text{where } C_y = C + C_T \quad \ldots [167]$$

$$\therefore V_o = \frac{1}{\beta} V_x = \frac{j\omega C_T R}{1 + j\omega C_f R} V \qquad \ldots [168]$$

and the magnitude and phase of the transfer can be calculated. This expression is further complicated by the finite open-loop gain and phase shift of the amplifier, so that all of these relationships must be considered approximate.

If R is very large, [166] becomes $|V_x| = \dfrac{1}{1 + C_x/C_T} V$

and [167] becomes

$$\frac{1}{\beta} = (C_f + C_y)/C_f$$

$$\therefore |V_o| = \frac{V}{1 + C_x/C_T} \times \frac{C_f + C_y}{C_f} = \frac{C_T}{C_f} V \qquad \ldots [169]$$

The measured quantity

$$Q = \frac{V C_T}{S}$$

$$= \frac{V_o}{S} C_f \qquad \ldots [170]$$

Comparing [170] with [164] we find that Q is independent of C and C_T, so that by using the circuit of Fig. 7.17(d) the cable capacity becomes unimportant. Such an arrangement is called a 'charge amplifier', since the output signal produced depends only on the charge produced by the transducer. Usually R is required to establish the d.c. operating condition for the amplifier, and its effect on bandwidth can be established from [168]. It also means that the transducer can be separated from the circuit by a suitable cable, and that changes of cable capacity with temperature will not have a significant effect on the output. Note that in [169] the magnitude of V_o is independent of frequency when $R \gg 1$.

7.1.8 Inductive sources

Many transducers are inductive by nature, e.g., differential transformers, variable reluctance, magnetic pick-offs, etc. In many cases
such a.c.-excited devices either produce a digital output, or have high-
level outputs which do not require amplification. However, in some
cases, for example microphones and record pick-ups, gain is required.
Consider Fig. 7.18(a). There are specially designed integrated circuits
for such applications (e.g. LM170), but we are here concerned with
using conventional op amps. R_L is the resistance of the inductive
source L.

$$G = \frac{R}{R_L + j\omega L} \qquad |G| = \frac{R}{\sqrt{R_L^2 + \omega^2 L^2}} \qquad \ldots [171]$$

$$\text{at} -3 \text{ dB}, \ |G| = \frac{1}{\sqrt{2}} \qquad \therefore f \text{ bandwidth} = \frac{\sqrt{2R^2 - R_L^2}}{2\pi L} \quad \ldots [172]$$

so that the gain is a complicated frequency-dependent expression.
This connection also introduces a phase lead within the loop, so that if
R_L is small compared with R, and C_s, the stray capacity across R_s, is
significant at high frequencies, there is a danger of circuit instability if
the amplifier internal lag compensation is inadequate. For the circuit
in Fig. 7.18(b) the gain is $(R_1 + R_2)/R_1$ and independent of L, so that
this is usually the best arrangement. However, if high-level input
signals are likely to be applied, for example in some audio applications, both input terminals of circuit (b) operate at the signal level,

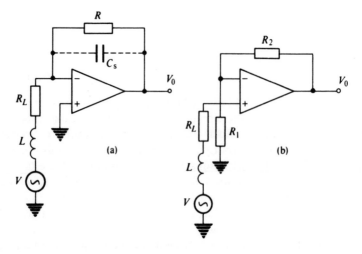

Fig. 7.18 Inductive sources

whereas the inputs of circuit (a) are at almost 0 V due to the virtual-earth action. This means that increased distortion can be expected from circuit (b) due to the non-infinite common-mode rejection ratio. It is likely to get worse with increasing frequency. The expressions of [171] and [172] are approximate, since for accurate calculations the amplifier open-loop gain and phase shift must be considered.

7.1.9 A.C. indicators

It is often desirable to build a simple circuit which will produce an output to drive an ordinary moving-coil meter, for example to monitor the output of a seismic mass self-generating vibration transducer. This type of transducer operates above its own damped resonant frequency, and is an a.c. generator since its output falls rapidly below resonance. The output is proportional to velocity, so that if displacement indications are required, an integration process must be introduced. What is required, then, is an a.c. amplifier, an integrator and a rectifier for driving the meter. Such an arrangement will be peak-reading, but the meter can be calibrated for the average or rms of a sinusoid. If the waveforms to be monitored are complex, then a true rms converter may be required. A possible circuit is shown in Fig. 7.19. A rigorous analysis is rather complex, and computer-aided techniques can be used to advantage. However, by a rational selection of component values approximations can be made.

The values of C_1 and C_4, both d.c. blocking components, are chosen to have negligible impedance at the frequencies of interest, and R_2

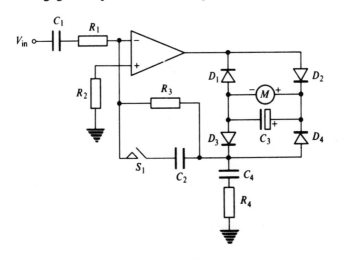

Fig. 7.19 A.C. indicator

should have the resistance 'seen' by the negative input terminal. For velocity indications, S_1 is open. On positive output signals, D_2 and D_3 conduct, and on negative-going signals D_1 and D_4 conduct. The current through M is therefore unipolar and full-wave rectified. The current in M is sensed by R_4 and a feedback voltage developed across it which is used as the loop control. In this way the meter is effectively driven from a constant current source, so that variations of meter-coil resistance (either with temperature or initial tolerance) do not affect the deflection. Providing $R_3 \gg R_4$,

$$\frac{V_{in}\mathrm{pk}}{R_1} = \frac{i\mathrm{dc}R_4}{R_3}$$

$$\therefore i\mathrm{dc} = \frac{R_3}{R_1 R_4} V_{in}\mathrm{pk}$$

The capacitor C_3 is used to smooth the rectified voltage appearing across the meter, and the smoothing time constant will be determined by C_3 and the meter-coil resistance. Minimisation of ripple is enhanced by increasing C_3 but response time will be increased, so there is a compromise between meter ripple at low frequencies and the speed of response to signal amplitude changes.

If S_1 is closed, the circuit now functions as an a.c. integrator in the way described for Fig. 7.21(b), and the output d.c. current is proportional to the source displacement. For a sinusoid displacement $x = A \sin \omega t$ and velocity $dx/dt = A\omega \cos \omega t$, where peak velocity is $A\omega$ and peak displacement is A.

7.2 Differentiation and integration

7.2.1 Differentiators

These two functions are readily performed with op amps, and have been used for many years (since the use of valve amlifiers) in analog computing networks. The tendency is for integrators to predominate, since the integrating process is used much more frequently in problem-solving in real situations and practical differentiator circuits can be more difficult to implement. The basic principle of the differentiator is shown in Fig. 7.20(a). The gain of the circuit can be written

$$G = \frac{R}{1/j\omega C} = j\omega CR, \text{ i.e. } \frac{V_o}{V_{in}} = -sCR \text{ and } V_o = -\frac{dV_{in}}{dt}CR$$

The product CR simply defines the magnitude of the differentiated

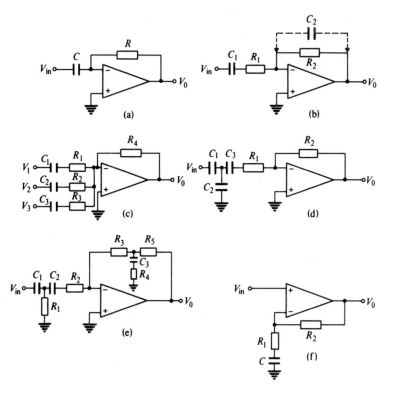

Fig. 7.20 Differentiators

waveform, e.g., if $V_{in} = kt^2$, then $V_o = -2KCRt$, a ramp produced from a squared function input, with the slope of the ramp $= -2KCR$. We must remember, however, that a single-pole internally compensated amplifier will introduce a phase lag of up to $90°$, and that the feedback loop as shown introduces another lag into the loop. This means that such a circuit is only marginally stable, and is therefore not to be recommended. Another difficulty with this circuit is that at high frequencies the gain approaches the amplifier open-loop gain. This means that noise and high-frequency radiated pick-up will be magnified and degrade the signal-to-noise ratio. In addition, the input impedance is low at high frequencies and will cause loading effects on the preceding circuit. A much more practical circuit is that shown in Fig. 7.20(b). The presence of R_1 limits the mid-frequency gain to $-R_2/R_1$ and increases the input resistance. It also limits the loop phase shift at low frequencies to

$$\tan \phi = \frac{-\omega C_1 R_2}{1 + \omega^2 C_1{}^2 R_1 (R_1 + R_2)} \qquad \dots [173]$$

in addition to the phase shift of the internal compensation. To increase the phase margin for frequencies beyond that of the required frequency range, C_2 can be added to produce a phase lead and further increase the stability margin. The phase lead inserted is given by

$$\tan \phi = \frac{\omega C_2 R_1{}^2}{R_1 + R_2 + \omega^2 C_2{}^2 R_2{}^2 R_1} \qquad \ldots [174]$$

The transfer function at low frequencies is given by

$$-V_o = \frac{\omega^2 C_1{}^2 R_1 R_2}{1 + \omega^2 C_1{}^2 R_1{}^2} + \frac{C_1 R_2}{1 + \omega^2 C_1{}^2 R_1{}^2} \frac{dV_{in}}{dt} \qquad \ldots [175]$$

From this expression the departure from the ideal differentiator which produces

$$V_o = -\frac{dV_{in}}{dt} C_1 R_2 \qquad \ldots [176]$$

when $\omega = 0$ in [175] can be obtained.

Various other useful functions can be produced by differentiator based circuits. The circuit in Fig. 7.20(c) is the basic extension which can be used for the sum of each differentiated input, so that

$$V_o = -R_4 \frac{d}{dt} (C_1 V_1 + C_2 V_2 + C_3 V_3) \qquad \ldots [177]$$

subject to the accuracy limitation expressed by [175]. It is interesting to note that many of the circuits used for integration can be applied in their differentiator 'dual' form, and this can be observed in Fig. 7.20(d). The equivalent integrator circuit uses resistors in place of capacitors, and a capacitor instead of R_2. Resistor R_1 is special to the differentiator for the reasons explained above. The basic purpose of this arrangement is to modify the relevant time constants without the normal constraints imposed by physical component values. A simple way to solve networks of this type is given in Appendix 14 and assumes that the amplifier input terminal is at virtual earth. It is shown that

$$\frac{V_o}{V_{in}} = \frac{-s C_1 R_2}{1 + (C_1 + C_2)/C_3 + s C_1 R_1 + s C_2 R_1} \qquad \ldots [178]$$

If $s R_1 (C_1 C_3 + C_2 C_3) \ll (C_1 + C_2 + C_3)$, we have the ideal function

$$V_o = \frac{-s C_1 C_3}{C_1 + C_2 + C_3} R_2 V_{in} = -\frac{dV_{in}}{dt} R_2 C_{eff} \qquad \ldots [179]$$

Where the effective value for $C_{eff} = C_1 C_3 / (C_1 + C_2 + C_3)$, thus producing a much smaller effective capacitor value and therefore a shorter time constant than would otherwise be practical. Other circuits exist for increasing the effective time constants,[75] and a more sophisticated technique for application to very low frequencies has been described by Bird.[111]

The circuit of Fig. 7.20(e) is interesting because it provides a method for finding the second derivative of a function using a single amplifier. The analysis is given in Appendix 15. Assuming that R_2 and R_4 can be made insignificant at the frequencies of interest, and that $C_3 = C_1 + C_2$ and $R_1 = R_3 R_5/(R_3 + R_5)$, then

$$V_o = -\frac{d^2 V_{in}}{dt^2} C_1 C_2 R_3 R_5 \qquad \dots [180]$$

and the time constants $C_1 R_3$, $C_2 R_5$ determine the output constant. When $C_1 = C_2$ and $R_3 = R_5$,

$$V_o = -\frac{d^2 V_{in}}{dt^2} CR$$

All of the above circuits are conventional inverting amplifiers, but the circuit of Fig. 7.20(f) is non-inverting. However, the function produced is not the direct derivative but consists of the input function added to its own derivative, as can be seen from the following:

$$V_o = \frac{1}{\beta} V_{in} = V_{in} \frac{1/sC + R_2}{1/sC}$$

$$\therefore V_o = V_{in}(1 + sCR_2) = V_{in} + \frac{dV_{in}}{dt} CR_2 \qquad \dots [181]$$

Clearly, the input function can be removed by using another amplifier, but then there is no component advantage compared with the use of circuit (b) and an inverter. However, the use of circuit (f) does provide a high input impedance circuit.

There are many other possible circuits for special applications, including bootstrapping techniques and differential input connections; the reader is referred to the literature for further study.

7.2.2 Integrators

Arguably the most important and most used of the fundamental op amp connections, integrators have been the subject of numerous developments, with the flexibility afforded by IC op amps allowing realisation of circuits which would have been uneconomic with discrete components. Figure 7.21 shows some basic circuits.

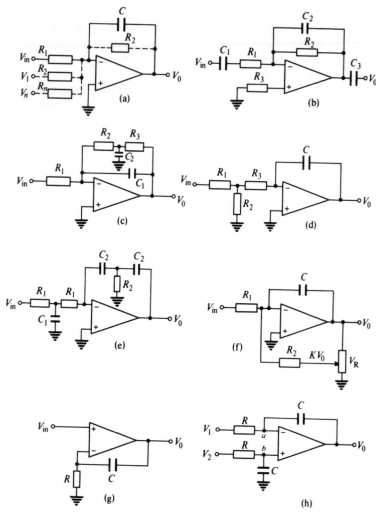

Fig. 7.21 Integrators

The ideal transfer characteristic for (a) would be

$$V_o = -\frac{V_{in}}{sCR_1} = -\frac{1}{CR_1} \int V_{in} \, dt$$

without R_2. If the integration of a steady-state or unipolar function is required, the integration process can only continue until the amplifier output saturates. In order to use the process again, initial conditions must be restored by removing the charge from C by discharging through R_2. Consequently, a reset switch is often necessary, connected in series with R_2. For example, if $V_{in} = Kt$, a ramp, $V_o = -Kt^2/2CR_1$, i.e., a squared function with the constant $-K/2CR_1$. The time constant CR_1 can be changed according to the required rate of change of signal, and it is possible to achieve very long time constants with the use of FET input amplifiers. The simple circuit can be extended in just the same way as the basic summing op amp to produce the integral of the sum of each input, and R_n can have differing values if required.

The basic operation is very simple; however, what we really need to know is how accurate can we expect the result to be. In many applications we must consider the capacitor leakage current (by drawing R_2 permanently in the circuit), and the temperature co-efficients of capacitance and resistance. These effects can be easily calculated, but we will assume here that we have 'perfect' components. The other main sources of error are the input offset voltage and current drift, and the finite open-loop gain of the amplifier. Notice that the feedback loop introduces a phase advance with an ultimate maximum of $+90°$, so that with internally compensated single-pole amplifiers there is no danger of instability, since the total loop phase shift varies from $-90°$ at very low frequencies to $-270°$ at high frequency, which is not the case for a differentiator. Another possible source of error is the open-loop input resistance of the amplifier, and the equivalent circuit for calculations of deviation from linearity is shown in Fig. 7.22. An analysis of this circuit leads to the expression:

$$G = \frac{-V_o}{V_{in}} = -R_{in}/[(R_1 + R_{in})/A + sCR_1R_{in}(1 + 1/A)] \qquad \dots [182]$$

If A is large,

$$G = -\frac{1}{sCR_1} \qquad \dots [183]$$

the basic integrator relationship. It is possible to calculate the errors

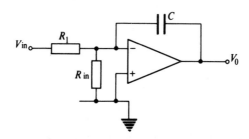

Fig. 7.22 Input resistance effect

due to finite A and R_{in} which cause the output to deviate from the ideal using [182] and [183].

If the input is a step function $E(H)t$, then the ideal output is a ramp. For example, if $C = 1\,\mu\text{F}$ and $R_1 = 1\,\text{M}\Omega$, the output slope is E V/sec. The response of the circuit to the complete expression [182] can be obtained by using the Laplace transform method and multiplying the input transform by the transfer function, i.e., $\mathscr{L}$ (output) $= 1/s \times$ equation [182] (see [186] to [189]). The resulting expression allows us to calculate the deviation from linearity of the output as a result of finite A and R_{in} when compared with the ideal output. This process is not restricted to step inputs, and the integration errors associated with any input waveform can be obtained by using the correct transform.

It should be noted that the foregoing assumes a simple real value for A, whereas in practice an internally compensated op amp will have almost $-90°$ phase shift associated with the single-pole roll-off at frequencies above a few cycles per second. Consequently, a rigorous analysis requires that the function $A_{(s)} = A_0/(1 + sT_0)$ be substituted for A in [182], where A_0 is the d.c. open-loop gain and $T_0 = 1/\omega_0$, the compensation breakpoint. This comment applies to any of the op amp analyses given in this text, but in many cases the A_0 value is so high that errors associated with it, even at frequencies well above ω_0, are often insignificant compared with other practical sources of error. Generally we can use [183] for low frequencies when A_0 is very large, but for calculating transient-response errors the complex function $A_{(s)}$ should be used in [182]. Simplified expressions have been derived by Stata[112] for demonstrating the sort of error to be expected.

The largest source of integration errors, especially at low frequencies, is usually the amplifier input voltage and current drift. Fortunately, these are somewhat easier to assess than the gain and input resistance errors. The input errors are taken to include drift with temperature, time and power supply. The input offset voltage error simply appears as though a small d.c. voltage source is connected in series with the source signal, and the offset current flowing in the source resistance produces another error voltage which can be added to the offset voltage. For minimum error it can be seen that the source resistance in each input terminal must be carefully balanced. Now,

$$V_o = \frac{-1}{CR_1} \int V_{in}\, dt, \quad \therefore \frac{dV_o}{dt} = \frac{-V_{in}}{CR_1} \qquad \dots [184]$$

With zero source signal, $V_{in} = V_{OS} + I_b R_1$ where V_{OS} is the input offset voltage and I_b the bias current. In order to substitute I_{OS}, offset current, for I_b, R_1 must be connected in *both* input terminals. Let us assume the input is balanced, so that

$$-\frac{dV_o}{dt} = \frac{V_{OS}}{CR_1} + \frac{I_{OS}R_1}{CR_1} = \frac{V_{OS}}{CR_1} + \frac{I_{OS}}{C} \qquad \ldots [185]$$

If we connect an integrator with balanced input terminals to 0 V and switch on, the output voltage changes in the way expressed by [185]. This indicates the limit of accuracy for low-frequency integration, and appears as a drift at the output. For a defined time constant CR_1, decreasing R_1 and increasing C minimises the drift contribution of I_{OS}. In practice there is a limit to which source resistance can be reduced and large capacitors with low leakage currents are very expensive.

The offset current can be minimised by the use of FET input amplifiers, but the temperature dependence of offset current must be carefully considered. Offset voltage errors are reduced by selecting pre-trimmed devices. Alternatively, one of the many offset balancing techniques described in section 4.2.1 can be applied to the integrator input. The resulting drift then depends on the degree of offset balance and its stability with temperature and time. Drift is inevitable, but by careful selection and design it can be reduced to very low levels. The introduction of R_2 (Fig. 7.21(a)) as a permanent connection will also help to reduce drift, and limits the output to a d.c. voltage of

$$V_o = \frac{-R_2}{R_1}(V_{OS} + I_{OS}R_1)$$

In some cases it is possible to introduce a resistor R_x into the non-inverting input terminal such that $V_{OS} = I_{OS}R_x$, but the drift with temperature may be degraded.

Many integrator applications do not require direct coupling, and so the a.c. integrator can be used to eliminate the drift problems of direct-coupled circuits. Figure 7.21(b) shows a typical arrangement and C_3 may be necessary if the d.c. output must be isolated. For high frequencies, when the impedances of C_1 and C_3 are negligible, the transfer function is given by

$$\frac{-V_o}{V_{in}} = \frac{R_2}{R_1(1 + sC_2R_2)} = \frac{1/C_2R_1}{1/C_2R_2 + s} \qquad \ldots [186]$$

The response to a step function input $E(H)t$ would be

$$\frac{1}{s}F = \frac{1/C_2R_1}{s(s + 1/C_2R_2)}$$

$$\therefore \text{ output} = \frac{R_2}{R_1}(1 - e^{-t/C_2R_2}) \qquad \ldots [187]$$

i.e. $\qquad \dfrac{-V_o}{V_{in}} = \dfrac{R_2}{R_1}\left(\dfrac{t}{C_2 R_2} - \dfrac{t^2}{2C_2{}^2 R_2{}^2} + \dfrac{t^3}{6C_2{}^3 R_2{}^3} - \cdots\right)$

$$\qquad\qquad\qquad\qquad\qquad\qquad\qquad\qquad\qquad \cdots [188]$$

$$= \dfrac{t}{R_1 C_2} - \dfrac{t^2}{2C_2{}^2 R_1 R_2} + \dfrac{t^3}{6C_2{}^3 R_1 R_2{}^2} \qquad \cdots [189]$$

If R_2 is infinite, [189] reverts to the basic integrator response. The simple departure from linearity can be easily obtained from [189]. The same process can be repeated for any type of input waveform in order to find the output response. Again we assume that the amplifier open-loop response does not have a significant effect on the result. The d.c. voltage produced at the amplifier output terminal by offset voltage and current is limited to:

$$V_O \approx \dfrac{R_2}{R_1}(V_{OS} + I_{OS} R_2) \qquad\qquad \cdots [190]$$

providing $R_3 = R_2$.

The circuit in Fig. 7.21(c) can be useful for integrating low-frequency a.c. waveforms. For accurate integration at low frequencies the value of R_2 in circuit (b) must be very large, and this leads to a high d.c. gain which may cause the amplifier output to saturate. For circuit (c) the d.c. gain is $(R_2 + R_3)/R_1$. At the frequencies of interest, C_2 is chosen to decouple the resistive feedback path and thus allow the impedance of C_1 to predominate in determining the transfer function. The relative component values can usually be determined more easily empirically – but for those interested in a rigorous analysis, the transfer function is given by:

$$-G = \dfrac{R_2 + R_3(1 + sC_2 R_2)}{R_1 + sC_1 R_1[R_2 + R_3(1 + sC_2 R_2)]} \qquad \cdots [191]$$

Circuit (d) can be used to achieve very long time constants when component limitations are restrictive, and effectively magnifies the integrated resistance value:

$$-G = \dfrac{R_2}{sC[R_1(R_2 + R_3) + R_2 R_3]} \qquad\qquad \cdots [192]$$

The effective time constant resistance obtained from [192] is $R_1 + R_3 + R_1 R_3/R_2$. It must be noted that the resistance in each amplifier input terminal must be balanced to minimise offset voltage and the errors of integration caused by it.

As has been mentioned previously (section 7.2.1), many basic circuit configurations can be used for integration and differentiation, but with C and R interchanged. An example of this is given in

Fig. 7.21(e), which shows a circuit to obtain double integration with one amplifier. It is the equivalent to the circuit analysed in Appendix 15, and the function realised is:

$$-G = \frac{1}{s^2 C_1 C_2 R_1 R_2} \qquad \ldots [193]$$

when $C_1 R_1 = 4C_2 R_2$.

Circuit (f) is an interesting variation of the basic integrator which provides flexibility in some applications. Assuming that V_R is a low resistance compared with R_2, the transfer function is given by:

$$-G = \frac{1/CR_1}{s + K/CR_2} \qquad \ldots [194]$$

The response to a step input, calculated as in [187] to [189], is

$$\frac{-V_o}{V_{in}} = \frac{t}{CR_1} - \frac{Kt^2}{2C^2 R_1 R_2} + \frac{K^2 t^3}{6C^3 R_1 R_2{}^2}$$

For small values of K, the output linearity can be adjusted.

The non-inverting configuration of circuit (g) provides a high input impedance, but as with the differentiator version the output is not a simple integral function, but gives the integral of the input plus its sum, viz:

$$+G = 1 + \frac{1}{sCR} \qquad \ldots [195]$$

A consideration of some of the many integrator circuits would not be informative without the inclusion of a differential input connection. Figure 7.21(h) shows the equivalent to a normal differential amplifier, and observation of the circuit leads us to expect that the output would be the integral of the input difference. Such an arrangement will be useful for operation where common-mode signals are present. A simple way to evaluate the result is to equate the voltages at a and b:

$$V_a = \frac{1}{1 + sCR} V_1 - \frac{sCR}{1 + sCR} V_o \qquad \ldots [196]$$

$$V_b = \frac{1}{1 + sCR} V_2 \qquad \ldots [197]$$

From [196] and [197] $\qquad V_a = V_b$

$$\therefore -V_o = \frac{V_2 - V_1}{sCR}$$

i.e., $$V_0 = -\frac{1}{CR} \int (V_2 - V_1)\,dt \qquad \ldots [198]$$

There are many other possible integrator-type configurations, and often improved performance is achieved by using several amplifiers to provide isolation between stages, higher open-loop gain, drift minimisation and slew-rate enhancement, etc. The basic principles applying to single-amplifier circuits have been outlined here, and the reader is encouraged to experiment with other networks as well as to consult the literature. Manufacturers' application notes are often invaluable sources, especially as new and improved ICs are introduced on to the market, since techniques which may have been rather impractical with earlier devices become valid alternatives. The situation is changing continuously, so that any practising engineer must read the technical journals in order to familiarise himself with market trends and device availability.

7.3 Voltage regulation

The vast choice of integrated regulator and reference devices currently available to the designer renders most regulator circuits based on conventional op amp techniques obsolete. However, op amps can still complement available devices, and a few alternatives are outlined here. The advance of switch-mode power supply circuits (SMPS) allows higher output powers to be provided whilst minimising internal dissipation; these evolved from designs using discrete op amps and transistors to the availability of special integrated circuits.[113-15]*

Among the special requirements which op-amp based regulators can provide are such parameters as very low temperature coefficients, low standby current, tracking outputs and increased power output. For many op amp circuit applications, reference supply lines can be provided as has been seen in section 4.5.

A basic regulator circuit may be drawn as in Fig. 7.23. There are a number of points to observe with such an arrangement. The unregulated input voltage is applied only to the output stage, with the error amplifier supply derived from the regulated output. This means that amplifier power-supply rejection ratio errors do not degrade the output performance. The reference device may well be so good that the constant current source I is not required, and the reference can be supplied directly from the output voltage by using a dropper resistor.

---

* e.g., Texas Instruments TL497

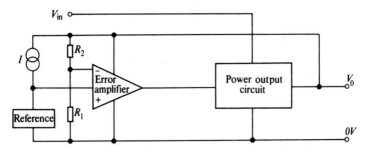

Fig. 7.23 Voltage regulator scheme

Alternatively, constant-current FET two-terminal devices may be used (e.g., Siliconix constant-current diodes). The resistors R_1 and R_2 are chosen to give the reference voltage at the inverting input with the desired output. This means that any output voltage errors (caused by variable loading on the output resistance or input voltage changes) are attenuated by $R_1/(R_1 + R_2)$ at the amplifier input; however, the open-loop amplifier gain can be made very high so that the loop gain of $AR_1/R_1 + R_2$ is still large.

Let $R_1/R_1 + R_2 = K_1$, and rejection of line voltage by the reference device $= K_2$. Then let the output voltage change by δV_o. The change at the amplifier input $\delta V_o(K_1 - K_2)$ produces an output of $A\delta V_o(K_1 - K_2)$ and produces an effective output change of $\delta V = \delta V_o[1 + A(K_1 - K_2)]$. Hence the ratio of the original change without feedback to change with feedback becomes $1/[1 + A(K_1 - K_2)]$, that is the regulation is determined by the loop gain. By selection of the correct amplifier, A can be very large even with a low supply voltage. The regulator output resistance is also reduced by the loop gain, and very low output resistances can be obtained. A very simple circuit, such as that shown in Fig. 7.24, is capable of producing high performance. The emitter-follower TR_1 output resistance (R_S/h_{fe}) is low even before the application of feedback. Only part of the input voltage change is reflected at the output before feedback, so the result is actually better than the loop-gain factor. If the amplifier positive supply terminal is returned to the input, Z_2 and R_4 are not required. This means that the input voltage is applied directly to the high-impedance collector, and the output change without feedback is even smaller; however, we now have the additional errors caused by the non-infinite value of the amplifier power-supply rejection ratio, again reduced by the loop gain. TR_2 protects the output from short-circuits, with the output current limited to about $0.6/R_6$. Note that the protection may be time-dependent, since the dissipation of TR_1 can be very high when the output is shorted to 0 V.

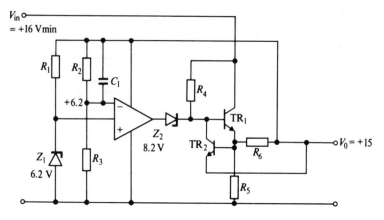

Fig. 7.24 Simple high-performance regulator

Another point to be aware of is that the output impedance of the circuit is likely to increase with frequency. This is because the amplifier open-loop gain (assuming single-pole compensation) decreases with frequency, and so also does the loop gain. Also, at very high frequencies, the current gain of TR_1 will decrease. It may be desirable to decouple the output terminal with a capacitor to reduce this effect. Capacitor C_1 helps to provide maximum a.c. feedback and improves stability margin.

If a simple zener reference Z_1 is used, the minimum temperature coefficient usually occurs between 5.6 and 6.2 V. If specified or selected low-drift devices are used, it is important to note that the best results are only obtained with constant-current biasing. This will not apply if V_o changes or if R_1 has a significant temperature coefficient of resistance.

The amplifier input terminals must be biased within the common-mode range; input offset drift causes output changes as a percentage of the reference voltage. Transistor TR_1 and Z_2 can be replaced by an 'N' channel JFET, the negative gate-to-source bias being used as the level-shifting method so that the amplifier output terminal operates within its range. Resistor R_4 will not be required since the gate of the FET does not require bias current like TR_1, and no zener current is needed. The main difficulty in using JFETs for this type of application is the very wide tolerance on operating characteristics typical with most devices.

Very flexible voltage references or regulators can be designed on the above lines for positive or negative supplies, and tracking dual-output circuits readily implemented on the basis of Fig. 2.21. The essential

advantages of providing any circuit with overall feedback are
apparent from the above discussion.

7.3.1 High-voltage floating regulators

The maximum operating voltage of most op amps is limited to less
than about ± 22 V, but a few devices are capable of more (e.g.,
National LM144 at ± 36 V and Burr-Brown 3584 at ± 150 V). Higher
voltages can often be obtained from hybrid devices, but they are
generally more expensive than monolithic amplifiers. One possible
solution to this problem is to 'float' the error amplifier at some level
above 0 V so that its maximum rating is not exceeded. A typical circuit
is shown in Fig. 7.25. A convenient way to understand how any circuit
operates is to write on to the circuit diagram typical operating
voltages as shown. R_1 and Z_1 provide basic regulation and voltage-
limiting, and amplifiers A_1 and A_2 float across the zener. The total
amplifier supply voltage is 33 V and the negative supply rail to each
amplifier floats on the output voltage. A_1 provides a stable reference
voltage of $+15$ V with respect to $V_{o/p}$, and the input terminals of A_1
are arranged to operate within the amplifier common-mode range. Z_2
provides level shifting so that the output of A_2 operates within its
range. The key to the circuit operation is the connection of $R_5 - R_8$.

Assume that the output voltage rises by $+\delta V_o$ as a result of a load
change. Since Ref and A_1 float on the output line, V_R must also change

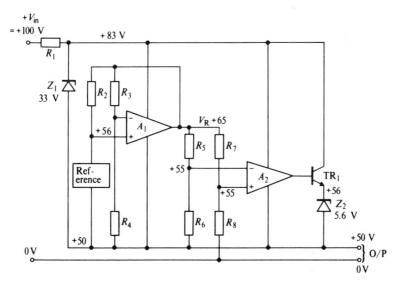

Fig. 7.25 High-voltage regulator

by $+\delta V_o$. The change at the inverting input of A_2 is $+\delta V_o$ since both ends of $R_5 - R_6$ change by $+\delta V_o$. Assuming the ratio of $R_7:R_8$ provides the voltages shown, the non-inverting input of A_2 changes by less than $+\delta V_O$, a value of $(55/65)\,\delta V_o$, i.e., $0.85\delta V_o$. This means that the overall feedback effect is negative, and the loop operates to reduce the original output voltage change. Although there is still a large proportion of δV_o applied to the non-inverting input, A_2 has a very high open-loop gain and the loop gain will also be high, thus producing good regulation. In this case, loop gain is $0.15\,A_2$. If the attenuation ratio $R_8/(R_7 + R_8)$ is maximised compatible with the required operating levels, then β, the effective feedback factor is increased $(\beta = 1 - R_8/(R_7 + R_8))$.

7.4 Sample-hold and peak-hold circuits

High-performance sample and hold circuits have been available in discrete modular and hybrid form for many years. More recently, monolithic devices have become available which require a minimum of external components. A typical example is the Burr-Brown SHC298. (It is likely that an external storage capacitor will always be required where hold times greater than a few milliseconds are necessary, due to the difficulty of fabricating large capacitors in integrated form.) These devices offer good performance and save expensive design time for the systems engineer. However, the flexibility and low cost of op-amp based circuits can still be useful, and many techniques have been suggested in the literature. A few of the basic circuits are included here as a background for more extensive study. Many of the requirements of sample and hold circuits are similar to those of peak hold; the latter are basically follower devices, where the memory retains the highest value of the analog signal at any one time, whereas the former store the signal amplitude at the instant of an external command. Among the more important parameters of these circuits are the time taken for the stored value to reach a defined percentage of the actual signal, and its rate of decay, frequently termed 'acquisition time' and 'droop' respectively.

Figure 7.26 shows some simple circuits. It is assumed in the following discussion that signals of either polarity can be handled by the necessary inversion of diodes or active devices. Circuit (a) is the simplest and most obvious starting point. Amplifier A_1 buffers the charge stored on C as a result of a switching signal applied to S. When J is turned on, the signal source charges C, with a speed dependent upon the 'on' resistance of J and the impedance and slewing rate of the source. When J is turned off, the charge on C leaks away through the

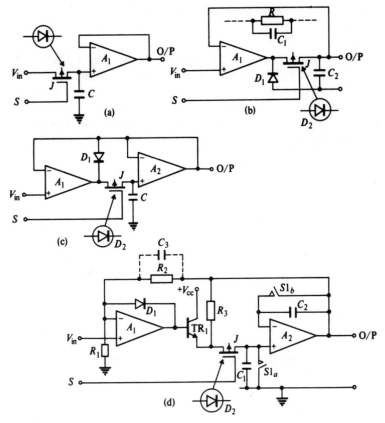

Fig. 7.26 Analog storage circuits

off-resistance and by supplying bias current to the amplifier. J may be either an MOS device or a JFET, but care must be taken to ensure that the correct switching voltage levels are applied to the gate in the presence of the maximum expected deviations of the signal, since the signal on the source and drain effectively change the gate-to-source bias. It is also possible to use integrated CMOS analog switches with their advantages of digital input control and nearly constant bi-directional 'on' resistance.

If $R = R_{on} + R_{source}$, then the acquisition time is determined by

$$V_o = V_{in}(1 - e^{-t/CR}) \qquad \qquad \dots [199]$$

and the percentage error is

$$\frac{V_{in} - V_o}{V_{in}} \times 100 \qquad \qquad \dots [200]$$

The percentage error from [199] and [200] is

$$100e^{-t/CR} \qquad \dots [201]$$

This represents the error at the capacitor, but we must add to this the effects of the buffer-amplifier rise time and slew rate. The time for which J must be 'on' for a given sample accuracy also depends on the rate of change of the source signal, since the signal itself will be changing during the acquisition period. When J is off, the charge on C will provide current through the leakage resistance of the switch and amplifier bias current so that

$$-C\frac{dV}{dt} = i_b + \frac{V}{R} \qquad \dots [202]$$

where V is the voltage on C at any time, R the leakage resistance and i_b the amplifier bias. Solving gives

$$V = i_b R(e^{-t/CR} - 1) + V_{in}e^{-t/CR} \qquad \dots [203]$$

where V_{in} is the initial voltage. The percentage error is given by $100 \times (1 - V/V_{in})$, and from [203],

$$\text{error} = 1 - e^{-t/CR}\left(\frac{i_b R}{V_{in}} + 1\right) + \frac{i_b R}{V_{in}} \qquad \dots [204]$$

The use of very low-leakage FET input amplifiers may render i_b negligible, in which case

$$\text{percentage error} = (1 - e^{-t/CR}) \times 100 \qquad \dots [205]$$

   This simple circuit will sample and hold signals of either polarity. A peak hold can be produced by connecting a diode in place of J, but the forward voltage drop of the diode limits the applicability. This problem is overcome by the application of feedback as shown in circuit (b). The diode D_1 prevents the amplifier output from saturating on negative-going signals, and a positive signal charges C_2 through D_2. The effective impedance and voltage drop of D_2 is now reduced by the loop gain so that it is very small. The acquisition time is limited by the amplifier slew rate and the maximum rate of change of output current (not necessarily identical). When the signal reduces below the stored level, D_2 is reverse-biased and the droop depends on the diode leakage and amplifier bias currents. The circuit will only have a low output resistance during the acquisition time, but if the load is constant the effects on droop can be calculated. A very high input resistance can be obtained, whereas circuit (a) has a low input impedance during the charging of C. The sample and hold can be produced by inserting J, and D_1 can be removed. One possible danger with this circuit is that of instability due to the fact that C_2 introduces another pole into the loop; however, this can be compensated by introducing a phase advance by connecting $C_1 R$ into the feedback path.

Circuit (c) requires two amplifiers but offers significant advantages when compared with single-amplifier arrangements. It is similar in operation to circuit (b), but A_2 buffers the storage capacitor and feedback is provided overall as well as locally. We now have high input impedance, low output impedance and low droop. Again the sample and hold and peak-hold functions for either polarity can be implemented.

Circuit (d) combines a number of further refinements to improve performance. Transistor TR_1 is included to improve rise time by providing a low-impedance source for charging C_1 and higher peak-current availability. The droop effect is minimised by the inclusion of C_2. The amplifier bias current generates an offset voltage by charging C_2 which is equal to the change of voltage on C_1 caused by bias current flowing in the non-inverting input. If the offset current is very low, the droop rate can be reduced very significantly. For best performance, the usual precautions against board leakage should be observed, and guard terminals utilised. The switches $S1_a$ and $S1_b$ can be used for resetting the peak hold when J is replaced by D_2. In the peak-hold application, diode D_2 must be a low-leakage device, and some minimisation of leakage can be obtained by connecting R_3 as shown. This has the effect of maintaining zero voltage across D_2 during the hold period, and thus minimising leakage current.

Another possible advantage is that the gain of the circuit can be made greater than unity. This is simply implemented by the presence of R_1 and R_2 so that $G = (R_1 + R_2)/R_1$. C_3 may be required to improve stability. Diode D_1 clamps the output of A_1 for negative transitions and prevents saturation. For negative peak hold the diodes are reversed and TR_1 becomes a pnp device. A circuit designed with this configuration using picoamp leakage diodes, and JFET amplifiers has been constructed by the author and can store 1 V to within 1 mV for more than 5 minutes at room temperature. It is important to select capacitors with the highest possible insulation resistance for C_1 and C_2, and ideally they should be matched in value.

8

Active filters

8.1 Some principles

The availability of low-cost op amps revolutionised the design of active filters, although many of the techniques had been established with discrete components and even valve circuits. The steady improvement of amplifiers – higher input impedance, greater open-loop gain, ease of frequency compensation, etc. – has offered great flexibility in implementing designs which may have been complex and expensive in discrete form. The abundance of information is demonstrated by the list of references on filter design at the end of this chapter – and this represents a very small selection of work which has been done. The advent of the quad op amp extends further the alternatives open to the designer, especially since the work of Kerwin et al.[F1]* on the state-variable principle. The use of multiple-amplifier circuits which might have been somewhat extravagant with single devices becomes much more attractive with multi-device packaging. In this chapter it is intended to discuss some basic ideas and to give some examples of circuits which can be easily implemented by the non-specialist user.

When an engineer is faced with a filter-selection problem for the first time, the choice available to him appears formidable. He may require to optimise any one, or a combination of several different parameters, and it is very difficult for him to decide which compromise best suits his needs; the key word here is compromise. It is true, of course, that computer-aided techniques are invaluable for fixing design values, but a lot of time can be wasted (and, incidentally, expense incurred) by using the computer to assist in decision-making, where a few minutes' reasoning will set the designer on to the right course from the start. It is clear that any optimisation process will

---

* F indicates references in the active filter reference list in section 8.6.

inevitably improve one aspect of performance to the degradation of another, and the filter type selected for any given application must tolerate this situation.

It is essential, then, that the initial approach must be a carefully thought-out definition of requirement. It is always true that no development task can be adequately controlled or processed in the correct manner unless a good specification document is assembled as the design objective; no less a criterion applies to the logical selection of a filter. Which then, are the most important parameters to be defined? There is bound to be a divergence of opinion here, but in the author's experience, the following list is likely to be applicable to the majority of users. Apart from the filter characteristics there are, of course, the major categories of:

1 Lowpass
2 Highpass
3 Bandpass ⎤
            ⎬ (and 'notch' filters)
4 Bandreject ⎦

The decision here is taken for granted. The major parameters to be considered are

1 Cut-off frequency ($-3\,$dB) or corner frequency (frequency at which the phase shift is $45\,°$).
2 Rate of attenuation beyond cut-off.
3 Phase response (or damping).
4 Transient response (rise time, overshoot, etc).
5 Passband ripple.
6 'Q' factor.

Since we shall be considering the implementation of active filters, it is assumed that the 'characteristic impedance' commonly detailed for passive networks will be replaced by the input and output impedance of the active circuit. These impedances are easily determined and can be modified by relevant buffer circuits, and they will not be considered further here. 'Group delay' is another parameter, which refers to the delay incurred by phase-shifting complex waveforms; it is more generally applicable to delay lines and lumped-constant prototype section filters.

There are obviously many differing techniques for implementing the various filter sections, and the object here is to establish a simple comparison between filters, so that the best choice for a given application is readily achieved. This comparison is based on the type of characteristic, not the means by which it can be achieved. There are many methods of implementation, including circuits such as the well-known Sallen and Key[F2] and Rauch networks, [F3] negative impedance converters[F4] and gyrators.[F5]

One of the most useful techniques for obtaining an appreciation of filter performance is that of the pole-zero plot. A pole is defined as a root of the denominator of the transfer function, and a zero as a root of the numerator. For example, the transfer function of a single R–C section lowpass filter can be written

$$T_1 = \frac{1}{1 + as} \qquad \qquad \dots [206]$$

and the pole is $-(1/a)$; the transfer function of a two-section R–C would be:

$$T_2 = \frac{1}{1 + as + bs^2}$$

and the poles would be

$$s = \frac{-a}{2b} \pm \sqrt{\frac{a^2}{4b^2} - \frac{1}{b}}$$

Hence any cascaded series can be expressed as the number of poles or zeros it contains; moreover, the position of the poles and zeros on the complex plane enables a prediction of the filter response to be made. The generalised approach is described by Sallen and Key[F2], where they say: 'Generally speaking, it is possible to select an appropriate network and a series of constants so that the poles and zeros can be placed anywhere in the complex plane. Under certain conditions all transfer functions of a given degree can be achieved with one fixed network, by selection of the appropriate active element constants.' So we see that, although the mathematical representation of the filter types such as Butterworth and Tchebyscheff are quite distinct polynomials, the transition between these types in terms of circuit implementation is not abrupt but a slowly changing form of response as the constants vary. Hence there is no reason why these specified sections must be used if the intermediate performance is required.

For example, then, a whole family of responses can be achieved between the maximally flat amplitude response of the Butterworth to the linear phase of the Bessel, allowing the compromise between good amplitude response and good transient response. This family of characteristics is called the transitional Butterworth–Thomson filters. [F6,F7] The realisation of such circuits is not limited to lumped-type networks, since these can be replaced by delay lines using a uniform structure or exponential taper. Such circuits have less sensitivity with respect to the active elements.[F6] A very good technique for designing the Bessel, Butterworth and Tchebyscheff filters up to 10 poles has been described by Shepard.[F8] The main families of characteristics will be briefly described below to consolidate the background. The descriptions will assume a 'lowpass' characteristic to simplify explanation but the arguments are equally applicable to 'highpass' filters.

8.1.1 Butterworth characteristic

These filters are described by a transfer function where the denominator is of a form known as the Butterworth polynomial, i.e.,

$$T = \frac{1}{D_{n(s)}} \qquad \ldots [207]$$

For example:

$$D_{6(s)} = s^6 + 3.864s^5 + 7.464s^4 + 9.141s^3 + 7.464s^2 + 3.864s + 1 \qquad \ldots [208]$$

and the roots can be shown to be:

$$s_{1,2} = -0.9659 \pm j0.2588$$
$$s_{3,4} = -0.7071 \pm j0.7071$$
$$s_{5,6} = -0.2588 \pm j0.9659$$

The positions of these on the complex plane can be seen to fall on the semicircle $0 - j$, -1, $0 + j$. These filters have the characteristic of being 'maximally flat' within the passband. This means that the amplitude-frequency plot is 'monotonic', that is, it is the flattest possible without changing direction, continuously increasing in attenuation as the frequency is increased. The roll-off beyond cut-off is also monotonic, eventually approaching 6 dB per octave for each pole. Hence a six-pole filter has a final slope of 36 dB per octave. The semicircle pole-location plots can be seen in Charts 8.1–8.3, where they are compared with the plots for other filter types. The Butterworth pole-location plot is the only one where all roots for increasing orders fall on the same line.

8.1.2 Bessel characteristic

Again the transfer function of the filter is described as above, but this time the denominator is in the form of a Bessel polynomial. The ideal linear phase-response filter can be characterised as a simple delay, and the usual way of expressing this delay is

$$T = \frac{1}{\cosh s + \sinh s} \qquad \dots [209]$$

$$= \frac{1/\sinh s}{1 + (\cosh s/\sinh s)}$$

An approximating polynomial can then be developed, and is known as the Bessel polynomial. The sixth-order polynomial, for example, becomes:

$$s^6 + 21s^5 + 210s^4 + 1260s^3 + 4725s^2 + 10\,395s + 10\,395$$
$$\dots [210]$$

Since the roots of this polynomial appear in the denominator of the lowpass transfer function, they are called poles as before. In the above example the roots, and hence the poles are located at:

$$s_{1,2} = -4.248 \pm j0.868$$
$$s_{3,4} = -2.516 \pm j4.493$$
$$s_{5,6} = -3.736 \pm j2.626$$

chart 8.1 shows the location of the poles compared with the Butterworth plot. The values of the roots quoted above have been shifted by a constant factor to facilitate comparison in the figure. (In some publications, 'normalised' values are quoted. This only amounts to a constant-factor shift, but when actually calculating component values the dependence of frequency on the normalised values must be carefully noted.)

The Bessel or linear phase filter is sometimes referred to as the 'maximally flat time delay' network. These filters have very little phase distortion within the passband, and are able to transmit the leading edges of pulses with very little overshoot or ringing. However, the price to be paid is that of a significant degradation of the frequency-response characteristic when compared with a Butterworth filter. It is typified by a gradual roll-off throughout the passband. The time delay is, of course, almost constant with frequency within the passband. These filters are well suited to voltage-to-frequency converters, pulse-modulation systems and applications requiring the accurate transmission of pulses.

8.1.3 *Tchebyscheff* (*characteristic*)[F5]

The transfer function is here characterised by a denominator having the form of a Tchebyscheff polynomial. This family of filters has a

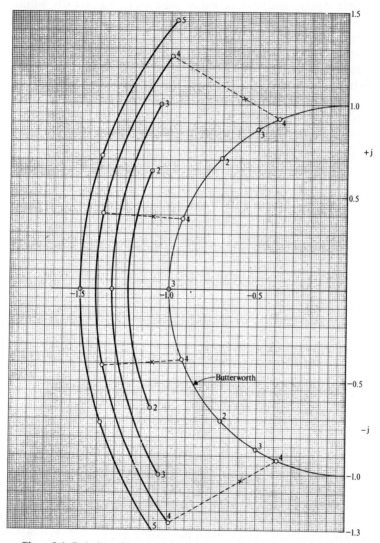

Chart 8.1 Pole location plots for Butterworth and Bessel functions.
Numbers are poles in function

ripple within the passband but the response curve turns over very rapidly into the constant rate of increased attenuation with frequency. The passband ripple can be controlled to have a defined magnitude by suitable design, and may be made quite small, e.g., 0.1 dB, whilst maintaining the rapid transition into the roll-off. Phase linearity and transient response is poor, and these filters tend to be used where selectivity is of greatest importance, for example in the separation of voice channels in data communication systems. Generally, the greater the passband ripple, the more rapid the rejection-band attenuation. The number of ripples in the passband is directly related to the number of poles in the transfer function, and the ripple is accomplished by arbitrarily assigning a constant multiplier to the polynomial. An example of the polynomial to the sixth order is:

$$32s^6 + 48s^4 + 18s^2 - 1 \qquad \ldots [211]$$

and the roots of this are (for a 2 dB ripple)

$$s_{1,2} = -0.174 \pm j0.261$$
$$s_{3,4} = -0.127 \pm j0.713$$
$$s_{5,6} = -0.0465 \pm j0.974$$

The amplitude response is then usually defined as

$$A(w) = \frac{1}{1 + E^2 T_n^{\,2}(w)} \qquad \ldots [212]$$

Where T_n is the nth-order polynomial and E^2 is the peak-to-peak error expressed by the passband ripple. Chart 8.2 gives examples of the pole-location plots compared with the Butterworth semicircle.

8.1.4 Legendre characteristic

This characteristic has a good attenuation performance with rather less passband ripple than the Tchebyscheff characteristic. Part of the band exhibits a linear phase characteristic, and is thus sometimes a reasonable trade-off between selectivity and ringing effects. Its performance is similar to that of a Tchebyscheff filter with a 0.1 dB passband ripple, and is monotonic outside the passband. A three-pole Legendre filter is equivalent to a four-pole Butterworth filter in terms of attenuation sharpness at cut-off, and a five-pole Legendre filter is similar to a nine-pole Butterworth filter. The Legendre polynomial appears to the sixth order as follows:

$$14.43s^6 - 19.69s^4 + 6.57s^2 - 0.312 \qquad \ldots [213]$$

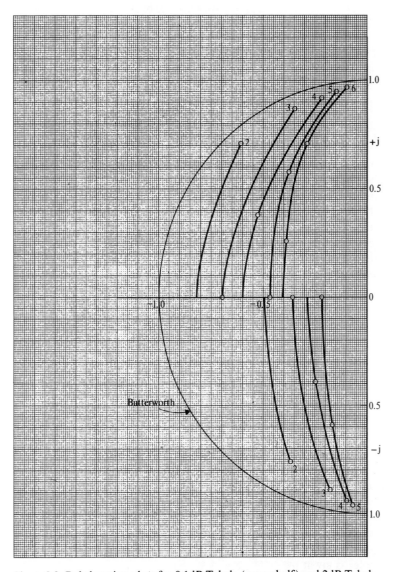

Chart 8.2 Pole location plots for 0.1dB Tcheb. (upper half) and 2dB Tcheb. (lower). Each plot is identical about the real axis.

and the roots of this are

$$s_{1,2} = -0.4390 \pm j0.2400$$
$$s_{3,4} = -0.3090 \pm j0.6980$$
$$s_{5,6} = -0.1152 \pm j0.9780$$

Chart 8.3 shows the pole plot for the functions to the 3rd, 4th, 5th and 6th orders.

8.1.5 Elliptic (Cauer) characteristic

Passband ripple is exhibited by a filter transfer characteristic formed from elliptic functions. Whereas the Tchebyscheff stop-band is monotonic, the elliptic filter stop-band consists of periodic variations between a predetermined minimum attenuation and a theoretical infinite attenuation. The number of passband ripples and also those in the stop-band again depend on the number of poles in the transfer function. One of the advantages of this type of filter is the very low insertion loss for the rate of attenuation above cut-off. The reader is referred to the literature for an exposition of the mathematics,[F9, F10, F11] but the basic functions are given by

$$\sqrt{P(w)} = \frac{w(w_1{}^2 - w^2) \cdots (w_k{}^2 - w^2)}{(1 - w_1{}^2 w^2) \cdots (1 - w_k{}^2 w^2)} \quad \text{(odd } n) \qquad \ldots [214]$$

$$\sqrt{P(w)} = \frac{(w_1{}^2 - w^2) \cdots (w_k{}^2 - w^2)}{(1 - w_1{}^2 w^2) \cdots (1 - w_k{}^2 w^2)} \quad \text{(even } n) \qquad \ldots [215]$$

8.1.6 Transitional Butterworth–Thomson characteristic

This type of characteristic has already been mentioned above. It represents a family of curves slowly changing from the Butterworth at one extreme to the Bessel at the other. Melsheimer[F12] describes in detail how the introduction of a constant factor into the transfer response of a Bessel characteristic, and the constant allowed to vary between 0 and 1, produces a family of curves with slowly changing properties between those of the Bessel and the Butterworth characteristics. This approach allows an accurate choice of the right curve for a specific application. The location of Butterworth–Thomson poles is given by

$$P_{BT} = R_T{}^m \underline{/\theta_B (m-1) - m\theta_T} \qquad \ldots [216]$$

where
$$R = \text{vector magnitude}$$
$$\theta = \text{vector angle}$$

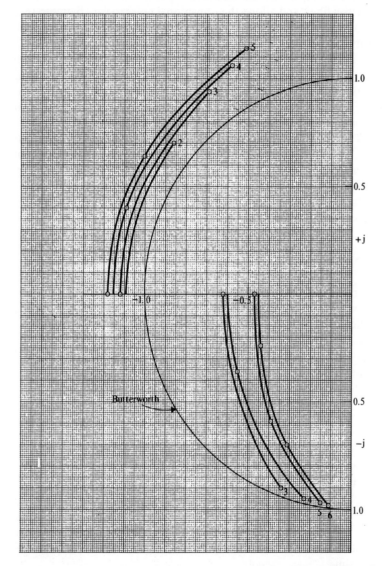

Chart 8.3 Pole location plots for Butterworth Thomson (Upper half) and Legendre (Lower). Each plot is identical about the real axis.

and m is allowed to vary between 0 and 1. This function is derived from the locus of the line joining each pole of a Butterworth to that of a Bessel filter, and allows the transitional poles to vary between these two limits, so that when $m = 0$ we have the pole locations for a Butterworth filter, $P_B = 1 \, \underline{/-\theta_B}$, and when $m = 1$ $P_T = R_T \, \underline{/-\theta_T}$ for the Thomson (Bessel) filter. For example, in Fig. 8.1 the pole locations for a fourth-order Butterworth and a fourth-order Bessel characteristic have been joined by dotted lines. The poles for transitional filters will lie along these lines, and one such example is given by the positions of the crosses.

This family of poles is plotted in Fig. 8.3. There is no reason why this procedure should not be extended further by moving pole locations beyond the Butterworth limit, thus enabling the designer to 'tailor' a filter to suit his own special requirement. Many possibilities present themselves, for example the same process could be used for producing responses occurring between any two of the other types of filter discussed here.

8.1.7 Optimum-L characteristic

A rather uncommon filter, the optimum-L, is typified by a response having less attenuation than either the Tchebyscheff or Legendre filters, but also less time delay over a greater bandwidth. As a result it has a better transient response than either of the other two. It has a monotonic amplitude response, and was originally proposed by Papoulis.[F13] The sixth-order polynomial in the transfer function is as below:

$$L_6(s) = 50s^{12} + 120s^{10} + 105s^8 + 40s^6 + 6s^4 + 1$$
$$\dots [217]$$

The roots and hence the poles can be solved from the factored form of this expression. For example, the third-order transfer function can be written

$$L_3(s) = \frac{0.577}{s^3 + 1.31s^2 + 1.359s + 0.577} \qquad \dots [218]$$

The factor 0.577 is used to produce unity d.c. gain. The poles then become

$$s_1 = -0.62 \pm j0$$
$$s_{2,3} = -0.345 \pm j0.901$$

The amplitude response is not maximally flat, but is monotonic, and has a sharper cut-off than the Butterworth characteristic.

8.1.8 Comparison of types

Generally, the linearity of the phase response is the factor determining the overshoot in the transient response. However, when the phase response is considered, it is usually within the context of the passband. Thus it is quite conceivable that a filter could have a reasonable phase linearity within the passband, but a very poor phase response outside of the passband. In this case, the response to transients having high harmonic content beyond the passband might be poor. Hence attention must always be paid to the out-of-band characteristics. The bandwidth itself is usually the most important factor determining the rise time. Table 8.1 gives a general comparison between the amplitude and phase responses. A good amplitude response is that which tends to rectangular and a good phase response that which produces a phase shift directly proportional to frequency. Clearly, there is no reason, apart from precedent and predictability, why one should be confined to the use of the known filter sections. For example, if the Tchebyscheff polynomial is arranged to produce a 0 dB ripple in the passband, the characteristic produced becomes a Butterworth.[F11] Another useful guide to selection of possible alternatives is by the use of the pole-location plots shown in Charts 8.1 to 8.3.

In each figure the pole-pair and the single-pole locations are plotted

Table 8.1 *Relative performance of different filter characteristics.*

Amplitude response	Filter type	Phase response
Worst	Bessel	Best
	Transitional Butterworth–Thompson	
	Butterworth	
	Optimum-*L*	
	Legendre	
	Tchebyscheff	
	0.1 dB ripple	
	n dB ripple	
	Elliptic (Cauer)	
	0.1 dB ripple	
Best	*n* dB ripple	Worst

relative to the plot for a Butterworth characteristic. The pole locations of increasing orders of Butterworth response all fall on the semicircle $0 + j, -1 + j0, 0 - j$. As we progress through Table 8.1 we see that the single-pole and pole-pair locations of successive types move from left to right across the plane. It is thus possible to observe the poles specific to a certain type of characteristic, and perform small modifications to filter performance by shifting the pole locations in the desired directions. There are many publications[F9, 10, 14, 15, 16, 17, 18, 19] in which characteristic curves of the various types of filter occur, and they will not be repeated here; rather, the reader is encouraged to refer to those references. Table 8.2 has been produced to afford a rapid comparison of the amplitude responses of the various types of filter.

The situation can rapidly become confusing, since it is common practice to define 'cut-off' as the -3 dB point of Bessel and Butterworth filters, whereas the 'cut-off' of a Tchebyscheff filter can be the n dB point of an n dB ripple filter.[F15] The 'cut-off' of a Bessel filter has also been defined as the frequency at which the phase shift is

Table 8.2 *Attenuation of different filters at frequencies relative to cut-off* (-3dB).

Response type	Number of poles	Approximate f/f_c for defined attenuation				
		0.1 dB	1 dB	3 dB	10 dB	20 dB
Butterworth...........	2	0.4	0.7	1	1.7	3.1
	3	0.53	0.8	1	1.46	2.2
	4	0.61	0.84	1	1.3	1.75
	5	0.66	0.88	1	1.23	1.6
	6	0.70	0.9	1	1.2	1.5
Tchebyscheff	2	0.55	0.75	1	1.6	3.0
	3	0.75	0.85	1	1.3	1.9
(0.1 dB ripple)	4	0.85	0.91	1	1.17	1.5
	5	0.89	0.93	1	1.11	1.32
	6	0.91	0.95	1	1.08	1.23
Bessel................	2	0.9	0.6	1	2	3.85
	3	0.9	0.6	1	1.75	2.8
	4	0.9	0.6	1	1.73	2.55
	5	0.9	0.6	1	1.73	2.4
	6	0.9	0.6	1	1.72	2.35
Legendre	2	0.42	0.7	1	1.75	3.1
	3	0.17	0.82	1	1.33	1.9
	4	0.16	0.89	1	1.21	1.55
	5	0.3	0.92	1	1.13	1.35
	6	0.3	0.95	1	1.1	1.25

Table 8.3 *Percentage overshoot to a step function*

Poles	Butterworth	Bessel	0.1 dB Tcheb.	Legendre
2	4.3	–	5.9	4.6
3	8.2	–	10.3	7.5
4	10.9	–	13.3	11.6
5	12.7	–	15.5	13.4
6	14.4	–	16.8	15.2

Table 8.4 *Relative rise time*

Poles	Butterworth	Bessel	0.1 dB Tcheb.	Legendre
1	1.0	1.0	1.0	1.0
2	0.97	0.95	0.96	1.04
3	1.05	0.98	1.07	1.13
4	1.11	1.0	1.16	1.21
5	1.15	1.0	1.27	1.31
6	1.2	1.0	1.36	1.39

Table 8.5 *Time to first peak (normalised to ω/ω_c)*

Poles	Butterworth	Bessel	0.1 dB Tcheb.	Legendre
2	4.4	–	4.5	4.3
3	4.9	–	5.2	5.1
4	5.6	–	6.2	6.1
5	6.4	–	7.3	7.2
6	7.1	–	8.3	8.2

one-half of the maximum.[F17] Tables 8.3–8.5 give comparisons of transient response. For an exposition of the transient response of Butterworth filters, the reader is referred to reference F20.

There are many other less well-known characteristics, such as the idealised Gaussian filter, linear phase with equi-ripple error filter, and transitional filters Gaussian over part of their range. The reader is referred to the literature for discussions of these, but it is hoped that this brief discussion will provide a background to the possibilities open to the engineer approaching this subject for the first time.

The outline of filter types described has been purely mathematical, but the very simple Sallen and Key circuit may be used to implement any of these characteristics simply by selecting component values to correspond with the necessary pole locations. This has been demonstrated by Shepard[F8] and has been used by the author for many applications. Typical circuit information is contained in the next section.

8.2 Some basic circuits

Figure 8.1 shows a simple implementation of some active filter circuits. Circuit (a) is a two-pole lowpass section, and (b) a three-pole, while (c) and (d) show the equivalent highpass sections. Each amplifier is connected in the unity-gain configuration, and filters with increasing numbers of poles may be cascaded, e.g., a four-pole is produced by a series connection of two cascaded two-poles, and a five-pole by a two- and three-pole circuit. This can be conveniently exploited as a result of the high input impedance and low output impedance of each section. Bandpass and band-reject characteristics can be obtained by the connections shown in circuits (e) and (f).

As an example of how to design a filter, let us assume that we have decided upon the need for a five-pole Tchebyscheff characteristic with 0.1 dB passband ripple to provide a lowpass function. The operating characteristics are: $f(-3 \text{ dB}) = 1$ kHz, ultimate rate of roll-off $= 30$ dB/8[ve] (100 dB/10[ade]). With reference to Fig.8.1(a) and (b), choose R to have a value compatible with the input requirements of the amplifier and the expected drift performance. Let $R = 10$ kΩ. Referring to tables,[F8] we find the following constants for the filter required:

5-pole 0.1 dB ripple Tchebyscheff characteristic

	C_1'	C_2'	C_3'
3-pole section	4.446	2.520	0.3804
2-pole section	6.810	0.1580	

Practical capacitor values are then obtained from the following expression:

$$C_1 = \frac{C_1'}{2\pi f R} \qquad \ldots [219]$$

where C_1' is the figure from the above table. For example,

$$C_2 \text{ (2-pole section)} = \frac{0.1580}{2\pi \, 10^3 \, 10^4} = 0.0025 \, \mu F$$

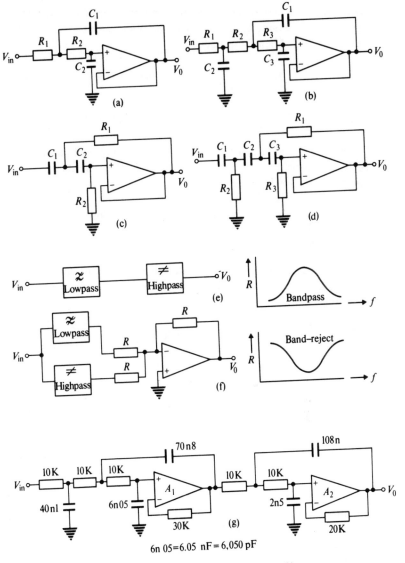

Fig. 8.1 Basic 'Sallen and Key' type filters

The completed circuit would appear as shown in Fig. 8.1(g).

If it is found that this process leads to inconvenient values for the capacitors, the resistor values can be varied; then from [219],

$$C_n = \frac{C_n'}{2\pi f K} \qquad \qquad \dots [220]$$

where K is the actual resistor value corresponding with the chosen capacitor. For example, in the above two-pole section, let $C_1 = 100n$ and $C_2 = 2n2$. Then from [220],

$$K_1 = \frac{C_1'}{2\pi \, 10^3 \, 10^{-7}} = 10.84 \text{ k}\Omega = R_1 \qquad \ldots [221]$$

and

$$K_2 = \frac{C_2'}{2\pi \, 10^3 \, 2.2 \times 10^{-9}} = 11.43 \text{ k}\Omega = R_2 \qquad \ldots [222]$$

After designing one or two filters using the above technique, it will be found to be a simple process to produce the desirable response curves. (N.B., for the three-pole filter solving for K_1 gives R_2, K_2 gives R_1, and K_3 gives R_3 in Fig. 8.1(b)).

The procedure for designing highpass filters is similar to the above, using circuits (c) and (d) as below:

1 Select type of filter.
2 Obtain table values, and calculate the reciprocal of each.
3 Choose the normalised capacitor value, e.g., $C = C_1 = C_2 = 10n$ in (c).

4 Calculate
$$R_1 = \frac{1}{2\pi f C C_1'} \qquad \ldots [223]$$

$$R_2 = \frac{1}{2\pi f C C_2'} \qquad \ldots [224]$$

where C_1' and C_2' are the normalised values from the table.

One of the principal advantages in the use of the filters described above is the fact that only unity-gain amplifiers are used, so there is very little chance of instability or oscillation. These circuits are, however, more susceptible to component tolerances than the state-variable type of filter described later (section 8.3). The bandpass circuit of Fig. 8.1(e) is not very good since at the peak the overall attenuation is equal to the combined loss of each section, and it may be necessary to restore that attenuation by a subsequent gain stage. An alternative bandpass filter circuit is described below.

8.2.1 Bandpass circuit

A very simple narrow-bandpass circuit is shown in Fig. 8.2.
This simple circuit has a number of advantages, viz.:

1 May be used for low- or high-Q applications.
2 Adjustment of R_2 provides 'notch' frequency change without affecting gain or bandwidth.

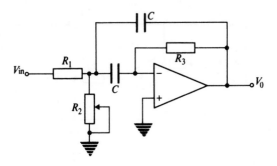

Fig. 8.2 Simple bandpass filter

3 Easy to adjust centre frequency gain.

The generalised transfer function is given by the following expression:

$$\frac{-V_o}{V_{in}} = \frac{s/R_1C_1}{s^2 + (s[C_1+C_2]/R_3C_1C_2) + ([R_1+R_2]/R_1R_2R_3C_1C_2)} \quad \dots [225]$$

In order to solve the expression it is necessary to assume some relationships, and the method is described in reference F21. The following expressions can be derived:

Centre frequency, $\quad f = 1 \Big/ 2\pi C \sqrt{\dfrac{R_1 R_2}{R_1 + R_2} R_3} \qquad \dots [226]$

Bandwidth, $\qquad\qquad B = \dfrac{1}{\pi C R_3} \qquad\qquad \dots [227]$

Gain at f, $\qquad\qquad G = \dfrac{R_3}{2R_1} \qquad\qquad \dots [228]$

$$Q = \frac{f}{B} = \frac{1}{2}\sqrt{\frac{R_3}{R_1 R_2/(R_1 + R_2)}} \qquad \dots [229]$$

8.2.2 Band reject filter

The design for a rejection filter shown in Fig. 8.3 is due to Bainter [F22] and is very useful because the sharpness of the 'null' is a function of amplifier gain rather than the matching of the passive components typical of earlier designs. This means that the response is predictable even in the presence of drift or ageing of the passive components. The sharpness of the notch is determined by the value of R_5, and for the

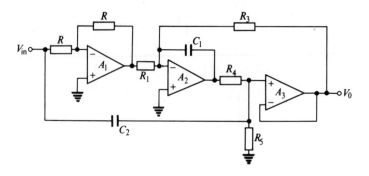

Fig. 8.3 Band reject filter

sharpest cut-off $R_5 = \infty$. The transfer function of this circuit is given by:

$$\frac{V_o}{V_{in}} = \frac{s^2 + (1/R_1R_4C_1C_2)}{s^2 + \left(s\left[R_4 + R_5\right]\Big/_2R_4R_5\right) + 1/R_3R_4C_1C_2} \qquad \dots [230]$$

When using R_5 it is good practice to make $R_4 = R_5$.

The notch frequency is

$$f = \frac{1}{2\pi \sqrt{R_1R_4C_1C_2}} \qquad \dots [231]$$

If $R_1 = R_3$, the d.c. gain of the circuit is unity, otherwise $G = R_3/R_1$. The source resistance of the signal V_{in} must be low, otherwise the gain will be affected. In applications where it is necessary to reject over a defined bandwidth rather than to remove an unwanted known frequency component, several notch sections each with differing frequencies f and differing widths (determined by R_5) may be cascaded.

8.3 State-variable filter[F1]

The principle of operation of this type of filter is based on the fact that any transfer-function polynomial can be factored into one or two degree parts. Each part is then realised by separate networks which are then cascaded to form the overall resultant. The great advantage of this approach is that the transfer function is very insensitive to the tolerances of the individual components when compared with the type of circuit described in section 8.2.1. High 'Q' values are obtainable without increasing the sensitivity to passive components.

Lowpass, highpass and bandpass characteristics can be obtained simultaneously, with band reject produced by summing the highpass and lowpass outputs. A very flexible filter 'building block' can be produced to offer many alternatives (e.g., Burr-Brown UAF 41 and National AF100). The advent of the quad op amp package made this approach even more attractive since it is a multi-amplifier technique. The basic circuit which produces a second-order response is shown in Fig. 8.4. The transfer functions to each output are as follows:

$$\frac{V_o}{V_{in}}(LP) = \frac{\dfrac{1}{R_4 R_6 C_1 C_2} \dfrac{1 + R_2/R_5}{1 + R_1/R_3}}{s^2 + s \dfrac{1}{R_4 C_1} \dfrac{1 + R_2/R_5}{1 + R_3/R_1} + \dfrac{R_2}{R_5} \dfrac{1}{R_4 R_6 C_1 C_2}} \qquad \ldots [232]$$

$$\frac{V_o}{V_{in}}(HP) = \frac{s^2 \dfrac{1 + R_2/R_5}{1 + R_1/R_3}}{s^2 + s \dfrac{1}{R_4 C_1} \dfrac{1 + R_2/R_5}{1 + R_3/R_1} + \dfrac{R_2}{R_5} \dfrac{1}{R_4 R_6 C_1 C_2}} \qquad \ldots [233]$$

$$\frac{V_o}{V_{in}}(BP) = \frac{-s \dfrac{1}{R_4 C_1} \dfrac{1 + R_2/R_5}{1 + R_1/R_3}}{s^2 + s \dfrac{1}{R_4 C_1} \dfrac{1 + R_2/R_5}{1 + R_3/R_1} + \dfrac{R_2}{R_5} \dfrac{1}{R_4 R_6 C_1 C_2}} \qquad \ldots [234]$$

A typical design requirement would be a two-pole lowpass Butterworth (maximally flat) characteristic. This can be realised as follows:

$$G, \text{ passband gain} = \frac{1 + R_5/R_2}{1 + R_1/R_3} \qquad \ldots [235]$$

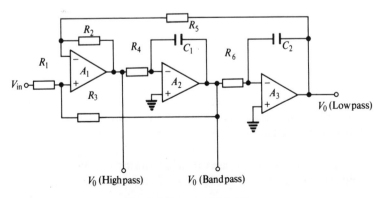

Fig. 8.4 State-variable filter

Let $R_1 = R_2 = R_5 = R_a$

$$\therefore G = \frac{2}{1 + R_a/R_3} \qquad \ldots [236]$$

Let $R_4 = R_6 = R_b$ and $C_1 = C_2 = C$. Then

$$f, \text{ cut-off frequency} = \frac{1}{2\pi C R_b} \qquad \ldots [237]$$

$$R_3 = \left(\frac{2}{\alpha} - 1\right) R_a \qquad \ldots [238]$$

$$G = 2 - \alpha \qquad \ldots [239]$$

where $\alpha = 1/\sqrt{2}$ for a two-pole Butterworth Filter. R_a, R_3 and R_1 may now be computed from [236], [238] and [239]. By varying α, different types of filter characteristic can be obtained. For further design information consult references 12, F23, F24.

The above procedure gives a two-pole active filter; a three-pole filter can be formed by connecting a single passive R–C network at the output (LP and HP), but the relative component values change since α is different. Higher-order filters are obtained by cascading the basic circuit.

8.4 Inductance simulation

The state-variable filter provides a low sensitivity to component variations, but in theory coupled L–C filters have the best performance in this respect.[F22] Since the separate sections of a state-variable filter built in cascade are uncoupled, there is a tendency for higher-order filters to be more susceptible to component variations than each individual stage. As a result there can be a requirement to produce simulated L–C-type filters, where the L-function can be produced from an active circuit. One such possible technique is the gyrator, a device which is now available in an IC package (e.g. Mullard TCA 580).

The ideal gyrator can be defined as a lossless two-port circuit that inverts a load impedance. If the load is a capacitor, then the gyrator input simulates an inductor. A basic gyrator circuit is shown in Fig. 8.5 (a), and the simple analysis can be evaluated as follows.

Assuming that A_1 and A_2 are very large, the input terminals of both amplifiers will be at the same level V_1. It follows that

$$V_3 = V_1 - i_1 R_1 = V_1 - i_2 R_2$$

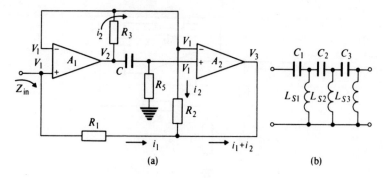

Fig. 8.5 Inductance simulation

$$\therefore i_2 = i_1 \frac{R_1}{R_2} \qquad \ldots [240]$$

$$V_2 = V_1 + i_2 R_3 \qquad \ldots [241]$$

$$V_1 = \frac{sCR_5}{1 + sCR_5} V_2 = \frac{sCR_5}{1 + sCR_5}(V_1 + i_2 R_3) \quad \ldots [242]$$

From [240] and [242],

$$V_1\left(1 - \frac{sCR_5}{1 + sCR_5}\right) = \frac{sCR_5}{(1 + sCR_5)} \frac{R_1}{R_2} R_3 i_1 \quad \ldots [243]$$

The equivalent input impedance is $Z_{in} = \dfrac{V_1}{i_1}$

Which from [243] is

$$Z_{in} = \frac{sCR_1 R_3 R_5}{R_2} \qquad \ldots [244]$$

Equation [244] thus has the impedance equivalent of an inductance whose value is

$$L_S = \frac{R_1 R_3 R_5}{R_2} C \qquad \ldots [245]$$

This circuit may now be used to simulate an inductor in an L–C-type filter network such as that shown in Fig. 8.5(b). In order to synthesise filter sections requiring 'floating' inductors, a 'floating' gyrator circuit can be developed from the basic circuit, and the reader is referred to the literature for a discussion of possible techniques.[F25]

Another simple but more limited method to simulate inductance is shown in Fig. 8.6.

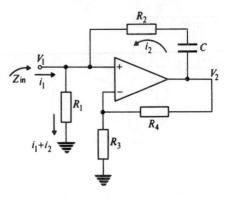

Fig. 8.6 Series L–R simulation

$$V_1 = R_1(i_1 + i_2)$$

$$\therefore i_2 = \frac{V_1}{R_1} - i_1 \qquad \qquad \dots [246]$$

$$V_2 - V_1 = i_2\left(R_2 + \frac{1}{sC}\right) \qquad \dots [247]$$

$$V_2 = \frac{R_3 + R_4}{R_3}V_1 = \left(1 + \frac{R_4}{R_3}\right)V_1 \qquad \dots [248]$$

Combining these gives

$$V_1\left(-\frac{R_4}{R_3} + \frac{R_2}{R_1} + \frac{1}{sCR_1}\right) = i_1\left(R_2 + \frac{1}{sC}\right) \qquad \dots [249]$$

$$Z_{in} = \frac{V_1}{i_1} = \frac{R_1 + sCR_1R_2}{1 + sC(R_2 - R_1R_4/R_3)} \qquad \dots [250]$$

If $R_2/R_1 = R_4/R_3$ in [250],

$$Z_{in} = R_1 + sCR_1R_2 \qquad \dots [251]$$

Equation [251] represents an inductance of value $L_s = R_1R_2C$ in series with a resistance R_1.

8.5 Capacitance multiplication

High-value capacitors are expensive, physically large, and have high leakage currents. They are often needed for simple decoupling and filtering applications. The circuit of Fig. 8.7 allows the realisation of an effectively large value using a small component.

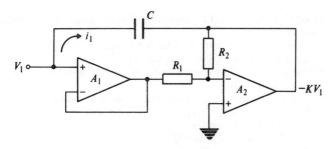

Fig. 8.7 Capacitance multiplier

$$i_1 = V_1 (1 + K) sC$$

$$\therefore Z_{in} = \frac{V_1}{i_1} = \frac{1}{(1 + K) sC} \qquad \dots [252]$$

The effective value for C is $(1 + K)C$, where $K = R_2/R_1$, and this value can be easily adjusted by changing R_2.

8.6 Some references on active filters

F1 W. J. Kerwin, L. P. Huelsman and R. W. Newcomb, 'State variable synthesis for insensitive integrated circuit transfer functions' *IEEE J. Solid State Circuits*, **SC-2**, 3, Sept. 1967.

F2 R. P. Sallen and E. L. Key, 'A practical method of designing RC active filters', *IRE Trans Circuit Theory*, March 1955.

F3 M. H. Nichols, H. Myron and L. Rauch, *Radio Telemetry*, Wiley, 1956.

F4 L. de Pian and A. Meltzer, 'Active filters: Negative impedance converters, Parts 3 and 4', *Electronics*, Sept. 2 and Sept. 16, 1968.

F5 L. de Pian, 'Active filters: Using the gyrator', Part 2; G. Aaronson, 'Synthetic Inductors from Gyrators', Part 10, *Electronics*, 10 June 1968, and 7 July 1969.

F6 M. N. S. Swamy and J. Walsh, 'Simple design procedure for active low pass filters using exponential R–C lines', *Proc IEE*, **120**, 1, Jan. 1973.

F7 Farouk Al-Nasser, 'Tables shorten design time for active filters', *Electronics*, 23 Oct. 1972.

F8 R. R. Shepard, 'Active filters: Part 12, Short cuts to network design', *Electronics*, 18 Aug. 1969.

F9 Anatol I. Zverev, *Handbook of Filter Synthesis*, Wiley, 1967.

F10 Y. J. Lubkin, *Filter Systems and Design*, Addison-Wesley, 1970.

F11 P. R. Geffe, *Simplified Modern Filter Design*, Rider, 1963.

F12 R. S. Melsheimer, 'If you need active filters', *Electronics Design*, 8, 12 April 1967.

F13 A. Papoulis, 'Optimum filters with monotonic responses', *Proc IRE*, **46**, 3, March 1958.

F14 'Special report on filters', *Electronic Products*, Sept. 1968.

F15 Russell Kincaid, 'Three pole active filter', *The Electronic Engineer*, June 1970.

F16 Einar C. Johnson, 'Take a short cut to filter design', *Electronics*, 25 Oct. 1971.

F17 *Catalog of Active Filters*, Burr-Brown Research Corporation, 1969.

F18 *Active Filter Modules*, Barr and Stroud Ltd.

F19 *Universal Active Filter Theory and Applications*, Kinetic Technology Incorporated, 1971.

F20 E. Van Vollenhoven *et al.*, 'Transient response of Butterworth filters', *IEEE Trans. Circuit Theory*, Dec. 1965, pp. 624–6.

F21 F. Stremler, 'Simple arithmetic: an easy way to design active bandpass filters', *Electronics*, 7 June 1971.

F22 J. R. Bainter, 'Active filter has stable notch, and response can be regulated', *Electronics*, 2 Oct. 1975.

F23 G. C. Temes and S. K. Mitra, *Modern Filter Theory and Design*, Wiley, 1973.

F24 R. W. Daniels *Approximation Methods for Electronic Filter Design*, McGraw-Hill, 1974.

F25 T. H. Lynch, 'The right gyrator trims the fat off active filters' *Electronics*, 21 July 1977.

F26 A. M. Hall and M. Holland, 'Filter design simplified with active filter modules', *Electronic Engineering*, Feb. 1972.

F27 B. A. Bowles and T. U. Nelson, 'Active filter design using IC gyrators', *Electronic Engineering*, Oct. 1977.

F28 A. Soliman and M. Fawzy, 'A universal active R-filter', *Electronic Engineering*, July 1977.

F29 W. J. Kerwin, L. P. Huelsman and R. W. Newcomb, 'State variable synthesis for insensitive integrated circuit transfer functions', *IEEE J. Solid State Circuits*, **SC-2**, 3, Sept. 1967.

F30 S. L. Guinta, 'Phase response characteristics of a Butterworth filter', *Application Report AR-100*, Krohn Hite Corporation, 1976.

F31 P. R. Geffe, 'How to build high quality filters out of low quality parts', *Electronics*, 11 Nov. 1976.

F32 S. Tintoprodjo and A. Van Meer, 'Forbidden regions for the poles of a third order equal valued resistor active lowpass filter', *Proc IEE*, **119**, 9, Sept. 1972.

F33 T. D. Reed, 'A simplified approach to the design of low pass active filters', *RAE Technical Report 67224*, Sept. 1967.

F34 K. Kraus, 'Bandpass/lowpass filter with two op amps', *Electronic Engineering*, Nov. 1975.

F35 D. J. Newport, 'An engineer's guide to Gaussian filters', *Electronic Engineering*, August 1975.

F36 'Amplifier active filter', *Electronics Industry*, Sept. 1975.

F37 M. Baril, 'Three mode network makes filter or oscillator', *Electronics*, 12 April 1973.

F38 'Electronics circuit aspects of active filters', *IEE Colloquium*, 15/1972.

F39 P. Cushing, 'Tunable active filter has switchable response', *Electronics*, 4 Jan. 1973.

F40 'Active filters with gain', *Fairchild Application Brief 132*, 1969.

F41 'RC active filters using integrated operational amplifiers', *S. G. S. Fairchild Application Report 162*, 1967.

F42 D. S. Miller, 'Susceptibility of active filters', *Electronic Components*, Oct. 1969.

F43 R. Jeffers and D. G. Haigh, 'Active RC lowpass filters for f.d.m. and p.c.m. systems', *Proc IEE*, **120**, 9, Sept. 1973.

F44 M. Artusy, 'Tunable active filter has controllable high Q', *Electronics*, 31 Jan. 1972.

F45 F. Ellem and D. Hazony, 'Active notch filters for eliminating noise', *Electronic Engineering*, Sept. 1972.

F46 N. B. Rowe, 'Designing a low-frequency active notch filter', *Electronic Engineering*, April 1972.

F47 D. Dekold, 'Wien bridge in notch filter gives 60 dB rejection', *Electronics*, 28 Aug. 1972.

F48 K. G. Beauchamp, 'A twin T filter design having an adjustable centre frequency', *Electronic Engineering*, June 1967.

F49 H. Haussmann and M. Kuo, 'Variable filter characteristics with an op amp', *Electronic Engineering*, Feb. 1972.

F50 R. Pease, 'An easily tunable notch pass filter', *Electronic Engineering*, Dec. 1971.

F51 P. Geffe, 'Bandpass filter shapes up from a low pass network', *Electronics*, 6 July 1970.

F52 R. Melen, 'Tunable active filter maintains constant Q', *Electronics*, 19 July 1971.

F53 R. Foss and B. Green, 'Design data for high and low pass active filters', *Plessey Co. Technical Communication*.

F54 *Handbook of Operational Amplifiers Active RC Networks*, Burr-Brown Corp., 1966.

F55 M. R. Lee, 'Active filters simplified', *Electronics*, 9 Nov. 1972.

F56 P. Ioannides, 'RC elements improve active filter performance', *Electronic Engineering*, May 1972.

F57 B. Gilbert, 'Nonlinear low pass filter rejects impulse signals', *Electronics*, 1 April 1976.

F58 I. Shepherd, 'Which filters?', *Electron* 21 Nov. 1974.

F59 G. Moschytz and R. Wyndrum, 'Active filters: Part 5. Applying the operational amplifier', *Electronics*, 9 Dec. 1968;
B. Welling, 'Part 6. The op amp saves time and money', *Electronics*, 3 Feb. 1969;
'Part 7. Analog blocks ensure stable design', 17 Feb. 1969.
'Part 8. Positive results from negative feedback', 31 March 1969.
V. Marsocci, 'Part 9. Applying nonlinear elements', 14 April 1969.
G. Aaronson, 'Part 10. Synthetic inductors from gyrators', 7 July 1969.
J. Mullaney, 'Part 11. Varying the approach', 21 July 1969.
M. Hills, 'Part 13. Narrowing the choice', 27 Oct. 1969.

F60 L. P. Huelsman (ed.), *Active Filters*, Dowden Hutchinson and Ross, 1976.

F61 Johnson and Hilburn, *Rapid Practical Designs of Active Filters*, Wiley, 1975.

F62 P. Bowron and F. W. Stephenson, *Active Filters for Communications and Instrumentation*, McGraw-Hill, 1979.

F63 Lam, *Analogue and Digital Filters*, Prentice-Hall, 1979.

F64 A. S. Sedra and P. O. Brackett, *Filter Theory and Design – Active and Passive*, Pitman, 1979.

F65 D. E. Johnson, J. R. Johnson and H. P. Moore, *A handbook of active filters*, Prentice-Hall, 1980.

9

Recent advances

It is inevitable that a chapter with this title will be out of date before it has been written. Therefore this recommendation is made to any practising circuit design engineer: no publication can ever relate the full story about integrated circuits and their availability since improvements in performance or device complexity are likely to continue for a long time. It is a fundamental necessity for the engineer to appraise books, journals, manufacturers' data and advertising on a continuing basis in order to remain up to date about devices available. As progress is made with device technology, so alternative circuit design techniques become practical, and reading the technical press should be a part of the engineer's work routine. In this chapter some of the more recent developments at the time of writing are mentioned to show the wealth of alternatives open to the linear-circuit designer.

9.1 Multi-amplifier packages

Two-, three- and four-amplifier packages are readily available; what, then, are the advantages to the user?

1. Since packaging of integrated circuits represents a large proportion of their total cost, it would be expected that cost per channel should be less. In a manufacturing environment, parts count, assembly and parts storing reductions helps to reduce costs. Reliability should be higher since interconnections can be reduced.
2. Complete families of same specification devices are available in 1-, 2-, 3- and 4-device packs.
3. Parameter matching such as input drift, noise, supply rejection and distortion enable the design of improved instrumentation amplifiers, audio amplifiers and active filters without device selection.

4. Space reduction is possible.
5. Universal biasing techniques can be used.

There are disadvantages. Crosstalk exists, and usually a specification is provided for this. It is usually worse at high frequencies, so beware of specifications where d.c. values are quoted and no curves or data are provided to indicate the degradation with frequency. In circuits where one amplifier is used for a switching function (e.g. a comparator) and another as a low-level amplifier, it may be advisable not to implement both functions within the same package. In addition to signal crosstalk, there are the effects of thermal crosstalk to consider. If one amplifier is driving a load which causes significant junction-temperature increases, the inputs of other amplifiers in the package may be subject to induced drift effects. The magnitude of the effect will depend on the physical layout of the chip (i.e., whether or not it is optimised for thermal feedback effects) as well as the actual specified input-stage drift performance.

Another limitation is power dissipation, since the same types of packages are often used for single and multiple devices, and it is usually the package type which limits the dissipation . This means that a quad amplifier will normally be specified for total dissipation, which limits each separate amplifier to one-quarter of the maximum rating.

Cost effectiveness is only likely to be optimised if all amplifiers in a quad package are used, although it might be just feasible to consider the availability of an uncommitted amplifier as a means of retrospective modification or as a spare in case of failure.

As the number of amplifiers in a single package is increased, so the available terminal pins decrease. Quad amplifiers do not in general have external balancing facilities available, and this may be a limitation in some applications. Some of the available quad op amps are listed in Table 9.1, using the commercial device specifications. Higher-performance devices are generally available at higher cost, and many manufacturers second-source the same basic device, sometimes with improved specifications. (Type numbers are often repeated, viz., RC3401 ≡ CA3401 ≡ MC3401.)

9.2 Current mirror ('Norton') inputs

There are a large number of amplifier applications which require many of the performance advantages of integrated op amps, but which do not demand the fully balanced differential input configuration of low-level, low-drift amplifiers. A family of op amps which uses the 'current mirror' technique has been developed to fulfil this

Table 9.1 Quad operational amplifiers

Type No.	Technology	Typical input offset voltage ($\pm$ mV)	Typical bias current (nA at 25° C)	Bandwidth at open loop gain = 1 (MHz)	Special features
LM324	Bipolar	2	−45	1	Single supply
LM346	Bipolar	2	−50	1	Programmable, low power
LF347	BIFET II	5	0.05	4	JFET input, low noise
LM348	Bipolar	1	30	1	Quad 741 low distortion
LM349	Bipolar	1	30	4(A = 5)	Quad 741 wideband
LM2902	Bipolar	2	−45	1	Automotive and industrial
LM3900	Bipolar	–	30	2.5	Current mirror input. Single supply.
TL064	BIFET	3	0.03	1	Low power
TL074	BIFET	3	0.03	3	Low noise audio, LM324 pin out
TL084	BIFET	5	0.03	3	General purpose
TL075	BIFET	3	0.03	3	RC4136 pinout
RC3401	Bipolar	–	50	4	Current mirror
RC3403A	Bipolar	2	−150	1	Ground sensing I/P low distortion
RC4136	Bipolar	0.5	−40	3	Low noise, wide band
HA4605	BIMOS	0.5	130	8	Low drift, wideband, low noise
HA4741	Bipolar	1	60	1	Quad 741

Device	Technology				Notes
XR4202	Bipolar	0.8	80	3.5	Programmable, low power
OP-09	Bipolar	0.6	300	2	Low noise, low drift.
OP-11	Bipolar	0.6	300	2	Matched, RC4136 pinout Quad 741. Low offset, LM348 pinout.
CA3401	Bipolar	–	50	5	Single supply, wideband
SA534	Bipolar	2	45	1	Single supply, low power
TDA0324D	Bipolar	2	45	1	Single supply, flat pack
MC3403	Bipolar	2	–200	1	Single supply, low distortion
MC3471	BIFET	6(max)	0.02	10	JFET input, wide band
MC4202C	Bipolar	1	200	2.5	Programmable, low power
µAF774C	BIFET	10(max)	0.2(max)	4	High slew rate wideband
ULN4236A	Bipolar	1	–30	1.5	Low power RC4136 pinout
ULN4436A	Bipolar	1	–30	1.5	Low power LM124 LM148 MC3403 pinout
RC4156	Bipolar	1	60	3.5	High slew, low noise
RC4157	Bipolar	1	60	19*	High speed
TAB1042	Bipolar	1	30	5*	Programmable
ICL7642B	CMOS	5(max)	0.001	1.4	Wide operating voltage very high input imp. Programmable power low bias
XR074C	BIFET	3	0.03	3	Low noise TL074 XR3403 pinout
XR084C	BIFET	5	0.03	3	TL084 LM324 pinout
TCA3002	Bipolar	0.8	–80	3.5	Programmable wide c.m. range
XR346	Bipolar	2	–50	1	Programmable

* Gain bandwidth product at $G = 5$

requirement. These amplifiers offer an alternative low-cost solution ideally suited to volume markets such as the automotive industry, and have been designed to operate from a single supply rail.[116] The second input to the amplifier is used to 'mirror' the non-inverting input current about ground and then to extract this current from that which is entering the inverting input terminal. When external resistors are connected in the input terminals, these currents are converted to voltages and the amplifier can then be used in the conventional manner of an op amp. Figure 9.1(a) shows the input stage arrangement. If i_b is very small, then β is approximately unity. Figure 9.1(b) shows how the amplifier is connected to operate as a single-supply a.c. amplifier. The equivalent circuit of this connection is shown in Fig. 9.1(c) to demonstrate how the amplifier is biased.

$$V_2 = i_1 R_2 + V d_1 \text{ and } i_1 = i_b + i_2$$
$$\therefore V_2 = V d_1 + R_2 (i_b + i_2) \qquad \ldots [253]$$

also $\qquad V_1 = i_2 R_3 + V d_2 \qquad \ldots [254]$

From [253] and [254],

$$V_2 = V d_1 + R_2 i_b + \frac{R_2}{R_3} V_1 - \frac{R_2}{R_3} V d_2 \qquad \ldots [255]$$

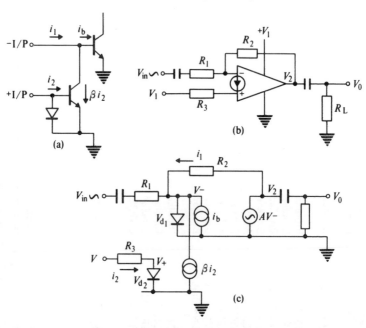

(a)

(b)

(c)

Fig. 9.1 Current-mirror input circuits

In practice $R_2 i_b$ is very small, therefore output voltage

$$V_2 = Vd_1 - \frac{R_2}{R_3} Vd_2 + \frac{R_2}{R_3} V_1 \qquad \ldots [256]$$

Normally V_1 is the positive power-supply voltage, so if $R_3 = 2R_2$, the output d.c. level will be about $V_1/2 - 0.3$ V. The a.c. closed-loop gain is $-R_2/R_1$. There are many ways of using the special features of this type of amplifier, and manufacturers' literature should be consulted for further design information. The basic structure of the circuit is relatively simple owing to the lack of necessity for a differential input stage and its associated common-mode rejection properties, so is well suited to low-cost quad amplifier configurations (e.g. CA3401, LM3900, MC3301).

9.3 Integrated-circuit chopper amplifiers

In section 5.13 various methods for reducing amplifier drift are discussed. One of the possibilities is the totally integrated chopper amplifier. Whilst very high performance in a very small volume is possible, these devices are complex chips, and this together with the low-volume market requirement is likely to cause their selling price to remain high. Table 9.2 compares typical specifications for four integrated-circuit differential-input chopper amplifiers. The input stability with time and temperature is excellent, but the penalty to be paid for this performance is the high noise level. It is pointed out for

Table 9.2 *Chopper amplifiers*

Parameter	Texas TL089	Datel AM-490-2	Harris HA2905	Teledyne Philbrick 1340
Input offset (μV)	25	20	20	20
Input bias (pA)	150	150	150	150
CMRR (dB)	120	120	160	160
Slew rate (V/μs)	10	2.5	2.5	2.5
Gain bandwidth (MHz)	3	3	3	3
Input resistance (MΩ)	100	100	100	100
Input offset drift (μV/$^\circ$C)	0.2	0.1	0.2	0.2
Input noise voltage (spot) (nV/$\sqrt{\text{Hz}}$ at 10 Hz)	–	–	900	–
Wideband total noise voltage 0.1 Hz to 10 Hz (μV$_{pp}$)	–	13	–	15

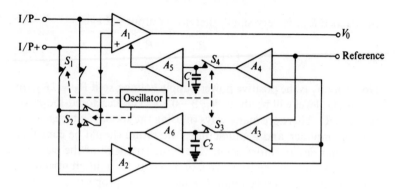

Fig. 9.2 Functional diagram of HA2905 integrated chopper amplifier

the HA2905 that chopper noise is present chiefly as a common-mode input current signal, and may be minimised by matching the impedances at the two inputs. Random noise may be reduced at the expense of bandwidth by using filters. Another consideration is the possible 'beat' effect caused by addition of signal frequency components near to the chopping frequency. The chopper frequency is given as 750 Hz for HA2905 and 1340, so if there is a possibility of signal frequencies occurring near to this, filters may be required.

The phase characteristic is very non-linear due to the internal separation of low frequencies and high frequencies within the amplifier. This means that for certain closed-loop gain values between open-loop and unity gain, the phase margin is decreased, and care should be taken to establish the necessary stability of the closed-loop gain at the value of interest. A functional diagram of the HA2905 is shown in Fig. 9.2.

9.4 The operational transconductance amplifier (OTA)

The OTA is a differential-input amplifier whose transfer characteristic is described in terms of an output current produced for a given input voltage rather than that of the conventional amplifier, which provides a voltage output. The basic transfer parameter is therefore transconductance, g_m, and is defined as $g_m = \delta i_0/\delta V_{in}$. The equivalent circuit can be drawn as shown in Fig. 9.3. The output voltage $V_o = i_0 R$, so that the voltage gain can be expressed as $G_V = g_m R$, where $R = R_o R_L/(R_o + R_L)$ when the load R_L is connected. Since these amplifiers are current-output devices, they would be expected to have a high output resistance, and so if they are to be used as voltage amplifiers a resistive load should be connected. The effective output resistance is reduced by the loop gain when feedback is applied.

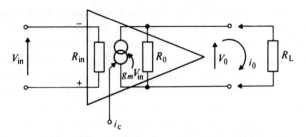

Fig. 9.3 Equivalent circuit of an operational transconductance amplifier

One example of an OTA is the CA3080 (RCA). This device is fabricated from transistors and diodes using 'current mirror' techniques. No resistors are used. In addition, it has an extra terminal, i_c in Fig. 9.3, which can be used to control the current bias of the differential-input stage, and thus modulate the value of g_m. This control terminal can be used as a gate terminal to switch the amplifier signal transfer on or off. It can therefore be used for multiplexing, sample and hold, and demodulation applications. The signal offset error (level shift) between input and output has a maximum of 5 mV. When the control terminal is used in the linear mode, the modification of g_m allows the device to be used as an amplitude modulator, automatic gain control or multiplying circuit. For further design-technique information see reference 117.

9.5 Micropower amplifiers

This term was originally applied to single op amps which could be operated with very low supply voltages and current. Typically, operation at ± 2 V and 20 μA and below is possible, and early devices include NE533, LM4250, μA735, RC4132 and μA776. High-value resistors are fabricated by depositing thin-film resistors on the surface of an integrated amplifier chip for control of very low currents. Some devices are 'programmable', i.e., their operating characteristics can be modified by external biasing. Later devices include triple (e.g. L144, ICL8023, WC788) and quad (e.g. XR4202, ICL7642) amplifiers. Individual amplifiers in multiple-device packages can be operated at very low powers, but since the supply rails are common to all circuits the total power may be higher than that for single devices. Between the limits set by traditional amplifiers and micropower types, there are the 'low-power' circuits such as LM346 which can operate at 350μA per amplifier.

There is usually a price to be paid for operation at very low power levels, so the designer should be very careful to ascertain the effects on circuit performance of such parameters as open-loop gain, input offset voltage, bandwidth, slew rate, and load-driving capability. One example of a programmable amplifier (HA2720) can be operated over a supply range of ± 1.2 to ± 18 V at currents of $1\,\mu$A to 1.5 mA and slew rates of 0.06 to 6 V/μS. This wide range of characteristics allows many different functions within a system to be implemented with a single device type, thus reducing documentation and the need for several replacement types. The OP–20 micropower amplifier offers a common-mode range which includes ground, and low input offset voltage drift.

9.6 Input-stage balancing

Chapter 5 describes how the differential input stage imbalance of both bipolar and FET input amplifiers affects the drift performance. However well designed and fabricated, input-stage devices will inevitably exhibit some degree of mismatch. This imbalance is often removed initially by external nulling. However, the external components themselves have temperature-dependent parameters and the method of application often affects the drift of the input stage. It is therefore desirable that the initial offset voltage is minimised at the chip level to reduce temperature dependence as much as possible. The reduction of initial offset also renders external nulling unnecessary in many applications and allows interchangeability. Two possible techniques used for balancing are 'zener zap'* and 'laser trimming' processes. Some devices with internal trimming are listed in Table 9.3.

9.6.1 Zener zap*

The equivalent circuit of the input stage is shown in Fig. 9.4. Assuming that successive stages are ideal and have zero offset, it can be shown that

$$\frac{R_{L2}}{R_{L1}} \times \frac{I_{S2}}{I_{S1}} = 1$$

when the offset voltage is zero, where I_S is the theoretical reverse saturation current of each transistor. The initial offset error is caused by processing variations producing differing I_S values for each device. The error can be nulled by adjusting R_{L1} or R_{L2} until the collector

---

* Precision Monolithics Inc.

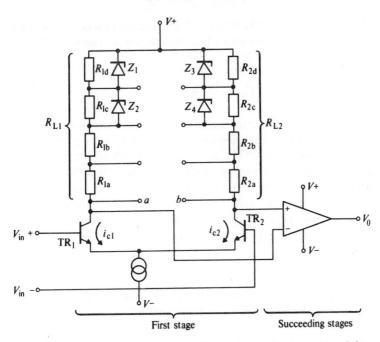

Fig. 9.4 'Zener zap' input-stage trimming (*Courtesy Precision Monolithics Inc.*)

Table 9.3 *Internally trimmed operational amplifiers*

Device type no.	Initial offset* voltage (μV) (typ)	Offset voltage* drift (μV/$^\circ$C) (typ)	Initial offset* current (nA) (typ)	Offset current* drift (pA/$^\circ$C) (typ)
OP-07D	60	0.7	0.8	12
OP-08G	250	1.5	0.08	1
OP-15G	500	4	0.005	100 pA max at 70°C
AD510J	100(max)	3(max)	5(max)	8 000 pA† (max) 0–70°C
AD517J	150(max)	3(max)	1(max)	1 500† pA (max) 0–70°C
BB510AM	150(max)	2(max)	20(max)	400(max)
BB3527AM	200	5	0.0003	10 at 70°C
μA714C	60	0.5	0.8	12
OP-20H	300	2.5	2	–

* Commercial specifications – improved devices available
† Total offset current

voltages are identical. The collector loads comprise a series of resistors R_{1a}, R_{1b}, etc., in parallel with which are integrated zeners Z_1, Z_2, etc. The offset voltage V_{OS} is given by

$$V_{OS} = \frac{KT}{q} \log_e \frac{R_{L2}}{R_{L1}} \cdot \frac{I_{S2}}{I_{S1}} \qquad \ldots [257]$$

The initial untrimmed error is measured, and then the necessary ratio R_{L2}/R_{L1} to null the error is recalled from a memory store consisting of values obtained from [257]. Current pulses of sufficient magnitude to fuse the required zeners are applied to the circuit so that the necessary parts of R_{L1} or R_{L2} are shorted to produce the correct ratio. Points *a* and *b* represent the usual connections for an external balance network. This process is carried out automatically on a production basis and provides high yields of very low offset devices.

9.6.2 Laser trimming

Thin-film resistors with their inherent low temperature coefficients are deposited on the silicon chip. The circuit is energised and the offset voltage monitored whilst a laser beam is used to selectively etch the appropriate resistors for minimisation of the error. Since the laser beam must produce a very high temperature at the chip to vaporise the resistor material, great care must be taken to ensure that the trimming process itself does not produce errors of such a magnitude as to swamp the error to be measured. Precautions include positioning resistors on the thermal 'centre line' of the chip, minimising the surface area to be heated and strobing of the output between successive laser pulses. These methods have been used for ICs in production and are described more fully in reference 118.

9.7 Single-supply, ground-sensing inputs

Traditionally, op amps have been used and specified for operation with dual polarity supplies, usually balanced about 0 V. This arises from the fact that the amplifiers have a common-mode input range limited in most cases to about 1–2 V less than the magnitude of either supply. This means that a conventional op amp operated from 0 V and a single positive supply must have its input terminals biased up from 0 V by 1–2 V for correct operation, and results in increased circuit complexity. In practice it is often more convenient to derive a second voltage from the single supply and then 'float' the amplifier at this new level in order to simulate a dual-supply configuration.

The availability of amplifiers with ground-sensing inputs has removed many of these earlier difficulties by permitting operation from a single supply with the input terminals operating at a common-mode level of 0 V. This performance is realised either by the use of pnp input transistors instead of the usual npn devices, or with FETs. A typical pnp input stage is shown in Fig. 9.5; the input bias current is sometimes specified as a negative value, since it flows out of, instead of into the base. Circuits using MOSFET input stages can operate below the 0 V line (e.g. CA3140).

Transistors TR_1 and TR_2 have their collectors connected to 0 V and operate with 0 V between base and collector. Since integrated pnp transistors tend to have lower current gain than npn types, the emitter-follower input connection minimises bias current. However, it does result in rather larger input offset drift figures due to the need for matching two sets of base-emitter voltages. The input devices operate with only 0.6 V collector-to-emitter voltage, so the saturation voltage must be low. The input terminals must be protected against the possibility of going more negative than about -0.3 V, otherwise the collector–base junctions become forward-biased and can be destroyed by high transient currents.

The output-stage design of single-supply amplifiers allows the signal swing to approach 0 V, but if it is necessary to sink output load current then the output saturation voltage rises accordingly. Note that in contrast with the input terminals, the output will never be at precisely 0 V, so that if the amplifier is used to drive some form of indicator it is not possible to set zero electrically. This is especially important where transducer sources are used and are likely to experience temperature-dependent zero shift. If the sense of the change is such that the output must change in a negative direction, a

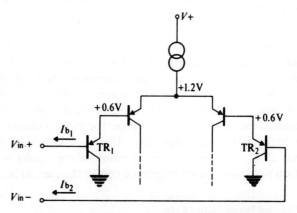

Fig. 9.5 Typical pnp ground-sensing input stage

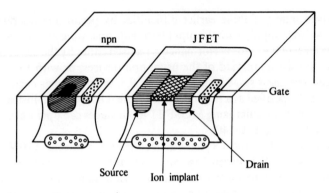

Fig. 9.6 BIFET technology (*Copyright National Semiconductor Corporation*)

dead band will be experienced under operating conditions due to the inability of the amplifier output terminal to traverse through 0 V.

9.8 BIFET* and BIMOS† amplifiers

These two terms are used to describe two processing techniques used in the fabrication of op amps. Until the appearance of these methods, most single-chip op amps were based on bipolar technology; hybrid circuits combined FET processing with bipolar chips to produce specialised circuits, e.g. JFET input amplifiers. These advances offer considerably extended flexibility in the optimisation of amplifier parameters.

The BIFET process enables well-matched high-voltage JFET devices to be incorporated on the same chip together with standard bipolar transistors. The basic construction is shown in Fig. 9.6.* The use of JFET input transistors provides low input bias current, high input resistance and low susceptibility to electrical overstress (compared with MOS devices), and allows replacement of more expensive hybrid or modular circuits with single-device convenience. The good matching characteristics of input pairs minimises offset voltage and drift, and the external balancing facility does not degrade drift as it does with some bipolar devices. The common-mode input range can be in excess of the positive supply voltage. The bipolar output stages provide high load-driving capability. The basic circuit simplicity minimises chip size, and the ease of manufacture results in high production yields and therefore reduced cost to the user. At least one

---

manufacturer offers a full range of single, dual and quad amplifiers in a variety of packages and with specially selected operating characteristics.* Input voltage parameters do not, however, compare favourably with the better bipolar input devices, which are usually more expensive.

The BIMOS amplifiers use bipolar and MOS devices on the same chip, and are claimed to be lower priced than BIFET (this situation may change, since pricing policy depends very much on yields, competition, market trends and negotiable terms). Again circuit simplicity and high yields keeps prices low, a typical amplifier requiring no more process steps than a conventional 741. The devices use gate-protected PMOS input transistors in a differential arrangement which is followed by a bipolar second stage. The output stage is formed by a complementary MOS pair. The input operating voltage extends below the negative supply rail, thus allowing ease of operation from single supplies. Open-loop voltage gain of a typical amplifier† is constant over a wide range of supply voltage, and the output terminal can swing to the rail in either direction. The MOS inputs provide a very high input impedance (up to $10^{12}\,\Omega$), but suffer a small offset shift with time under differential d.c. bias conditions.

An interesting comparison between bipolar and FET op amps is to be found in reference 119.

9.9 BIFET II ‡

Another process development subsequent to the early BIFET amplifiers has allowed fabrication of low-cost amplifiers with large power-bandwidth, low noise and good slew rate with low power consumption (e.g. LF351).[120] These amplifiers use a less complex and smaller circuit which is not optimised for minimum offset voltage. The JFET transistors are used only for the input stage, and initial offset trim is achieved internally by opening up links across binary weighted parts of resistors in the emitter and base of a current mirror used as the first-stage load. These amplifiers are well suited to audio applications, and can handle large input voltage swings with low distortion due to the low transconductance of the FET input.

---

* Texas Instruments Inc.
† CA3160, RCA Corp.
‡ © National Semiconductor Corp.

9.10 Triple passivation*

This is a proprietary process designed to improve reliability, long-term stability and reduce noise. The name 'triple passivation' originates from the fact that a three-step process is used to protect the surface of a chip against contamination. Thermally-grown silicon dioxide is initially deposited on the chip. This protects the junctions and acts as a 'getter' for impurities, but will not be a barrier against natrium. A second silicon nitride layer blocks natrium and is an additional protection against contamination. The third layer is a thick glass coating affording mechanical protection. Since the stability and drift with time of op amps is largely determined by chip surface contamination, it is claimed that this protection process improves performance.

9.11 Custom-designed integrated circuits

Most of the standard integrated circuits available to the designer are produced in very large quantities by the manufacturer. In this way the vast capital expenditure involved in the development and fabrication of circuits can be amortised and minimises the cost of devices. The price to the user is therefore very dependent on the achievement of high yields and automated production techniques. Many manufacturers diffuse the chips in one plant, and then ship them to areas of low labour cost for packaging, since the latter part of the process is more labour-intensive.

Nevertheless, it is still possible for a user to design a chip specifically for his own applications, and retain a circuit manufacturer to fabricate it. This technique is more typically orientated towards digital circuits, but is also possible with linear circuits. The design and manufacture of the diffusion masks is very expensive, and the initial investment in production renders this approach impractical for all but the very large-volume users. There are, however, some research organisations and very large companies who design and manufacture their own circuits for use in proprietary equipment.

The custom-design arrangement will probably continue to be more popular for digital rather than linear circuits, since many digital functions can be realised by selective connection of the same basically simple circuit cells (viz., FPLA – field programmable logic arrays and PROM). The gap for linear-circuit design is filled by the custom design of thick-film circuit modules (or possibly thin-film), and linear

* © Precision Monolithics Corp.

arrays such as the Ferranti 'Monochip'. In this latter device, arrays of up to 500 components including transistors, diodes and resistors, are interconnected by a metallised pattern to implement the customer circuit design.

This technique of producing linear circuits becomes economic at quantities as low as a few hundreds after the initial development cost has been met.

9.12 Low-voltage linear integrated circuits

Op amp capabilities are improving continuously, and advances of performance include low-power operation, high-power output, increased speed and input-error minimisation. Devices are available offering very good specifications at supply voltages down to 1 V total.

One such device is the LM10, and there are several sources describing this amplifier and its applications.[108,121-2] It consists of a voltage reference and an op amp on a single chip. It can operate in a floating mode from a single 1.1 V supply. This design produces a very flexible circuit which can be used for a wide variety of applications. The common-mode range includes the V-supply, and the output swings to within 50 mV of the supplies (depending on current output). It also has thermal overload protection. A wide variety of useful circuits can be constructed, ranging from microphone amplifiers to voltage regulators and signal transmitters.

Another very interesting device family is the ICL76XX range. These are all CMOS amplifiers and can be programmed to operate from currents of 10 μA to 1 mA, and with supply voltages of ± 0.5 V to ± 8 V. The input impedance is 10^{12} Ω and bias current typically 1 pA. One type (7612) has an input common-mode range that extends beyond either supply rail, and the output swings very near to the supplies. One possible disadvantage is that input offset voltage and drift is not quite as good as that obtained from bipolar devices.

9.13 Commutating auto-zero amplifier (caz)

This amplifier has been designed as an alternative to hybrid or monolithic amplifiers where very low input offset voltage and drift are required. The typical input offset is 2 μV and drift 0.005 μV/°C. The use of this device demands close attention to source thermoelectric effects and temperature-gradient minimisation to exploit its high performance. Input bias current is typically 300 pA and unity gain bandwidth 300 kHz at 1.7 mA supply current.

The basic block diagram is shown in Fig. 9.7. The on-chip

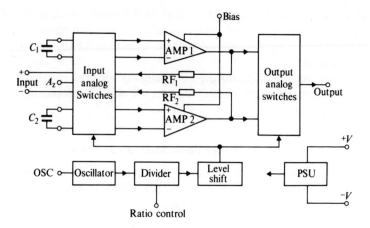

Fig. 9.7 ICL76XX block diagram (*Courtesy Intersil Inc.*)

oscillator, divider and level-shifting circuits generate the commutating signals used to operate the input and output switches in the correct sequence. One amplifier is used to charge a capacitor to the voltage which will subsequently null the input errors, whilst the other amplifier is connected to the signal. The situation is then reversed so that at any time one amplifier is connected to the signal in such a way that the previously stored value nulls the errors, whilst the other is storing its own error for application on the next cycle.

This amplifier is most suitable for application at low frequencies, since it is usually necessary to filter out the transients caused by the commutating switches.

Appendices

Appendix 1. Frequency and phase response of cascaded (buffered) R–C circuits

1.1 Single R–C

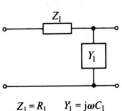

$$Z_1 = R_1 \qquad Y_1 = j\omega C_1$$

Fig. A.1

In the 'a' matrix form, the inverse of the voltage transfer function is given by
$a_{11} = 1 + Z_1 Y_1 = 1 + j\omega C_1 R_1$.
Hence the magnitude of the inverse transfer function $|a_{11}|$
$= \sqrt{1 + \omega^2 C_1{}^2 R_1{}^2} = \sqrt{1 + \omega^2 T_1{}^2}$
At the -3 dB frequency, $|a_{11}| = \sqrt{2}$

$$\therefore \omega_0{}^2 T_1{}^2 = 1. \qquad \therefore \omega_0{}^2 = \frac{1}{T_1{}^2}, \text{ where } \omega_0 = 2\pi f_0.$$

and $f_0 = -3$ dB frequency

$$\therefore \text{ Relative response } \left| \frac{1}{a_{11}} \right| = \frac{1}{\sqrt{1 + (f/f_0)^2}} \quad \text{where } f = \text{any frequency}$$
$$\dots [1]$$

Phase response is $\phi = -\tan^{-1}\omega T = \tan^{-1}(f/f_0)$ $\qquad \dots [2]$

1.2 Double (buffered) section

Let $C_1 R_1 = T_1$ and $C_2 R_2 = T_2$
For each section $a_{11} = 1 + ZY$

$$\therefore a_{11} \text{ for the combination} = (1 + j\omega T_1)(1 + j\omega T_2)$$

281

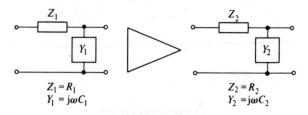

$Z_1 = R_1$
$Y_1 = j\omega C_1$

$Z_2 = R_2$
$Y_2 = j\omega C_2$

Fig. A.2

$$\therefore |a_{11}| = \sqrt{(1 - \omega^2 T_1 T_2)^2 + \omega^2 (T_1 + T_2)^2} = \sqrt{2} \text{ at } f_0$$

$$\therefore \omega_0^4 T_1^2 T_2^2 + \omega_0^2 (T_1^2 + T_2^2) - 1 = 0$$

Solving,
$$\omega_0^2 = -\frac{(T_1^2 + T_2^2)}{2 T_1^2 T_2^2} \pm \sqrt{\frac{(T_1^2 + T_2^2)^2 + 4 T_1^2 T_2^2}{4 T_1^4 T_2^4}}$$

In the simple case, where $T_1 = T_2$,

$$|a_{11}| = \sqrt{(1 - \omega^2 T^2)^2 + 4\omega^2 T^2} = \sqrt{2} \text{ at } f_0$$

$$\therefore \omega_0^4 T^4 + 2\omega_0^2 T^2 - 1 = 0$$

Solving,
$$\omega_0^2 = \frac{\sqrt{2} - 1}{T^2} = \frac{0.414}{T^2}$$

Relative response $\left| \dfrac{1}{a_{11}} \right| = \dfrac{1}{\sqrt{1 + 0.172 (f/f_0)^4 + 0.828 (f/f_0)^2}}$ . . . [3]

Phase response $\phi = -\tan^{-1} \left[\dfrac{2\omega T}{1 - \omega^2 T^2} \right]$

Relative phase response $\phi = -\tan^{-1} \left[\dfrac{1.287 (f/f_0)}{1 - 0.414 (f/f_0)^2} \right]$ . . . [4]

1.3 Three-stage (buffered) R–C

Let $T_1 = T_2 = T_3$ and $\omega T = x$.

Then $a_{11} = (1 + jx)^3 = (1 - 3x^2) + j(3x - x^3)$

$$|a_{11}| = \sqrt{1 + x^6 + 3x^4 + 3x^2} = \sqrt{2} \text{ at } f_0$$

This can be solved for x using Newton's method, and $x^2 = 0.26$.

$$\therefore \omega_0^2 = \frac{0.26}{T^2}$$

Relative response

$$\left| \frac{1}{a_{11}} \right| = \frac{1}{\sqrt{1 + 0.018 (f/f_0)^6 + 0.203 (f/f_0)^4 + 0.78 (f/f_0)^2}}$$ . . . [5]

Phase response

$$\phi = -\tan^{-1}\left[\frac{3\omega T - \omega^3 T^3}{1 - 3\omega^2 T^2}\right]$$

Relative phase response

$$\phi = -\tan^{-1}\left[\frac{1.53\,(f/f_0) - 0.133\,(f/f_0)^3}{1 - 0.78\,(f/f_0)^2}\right] \qquad \ldots [6]$$

1.4 Four-stage (buffered) R–C

$$a_{11} = (1 + jx)^4 = (1 + x^4 - 6x^2) + 4j(x - x^3)$$

$$|a_{11}| = \sqrt{(1 + x^8 + 4x^6 + 6x^4 + 4x^2)} = \sqrt{2} \text{ at } f_0$$

Solving $\qquad x^2 = 0.189 \qquad \therefore \omega_0{}^2 = \dfrac{0.189}{T^2}$

Relative response

$$\left|\frac{1}{a_{11}}\right| = \frac{1}{\sqrt{1 + 0.0013(f/f_0)^8 + 0.027(f/f_0)^6 + 0.215(f/f_0)^4 + 0.757(f/f_0)^2}} \qquad \ldots [7]$$

Phase response

$$\phi = -\tan^{-1}\left[\frac{4(\omega T - \omega^3 T^3)}{1 + \omega^4 T^4 - 6\omega^2 T^2}\right]$$

Relative phase response

$$\phi = -\tan^{-1}\left[\frac{1.74(f/f_0) - 0.329(f/f_0)^3}{1 + 0.036(f/f_0)^4 - 1.135(f/f_0)^2}\right] \qquad \ldots [8]$$

Appendix 2. Frequency and phase response of equal-valued R–C ladder

2.1 Single stage

The single stage response will be as for section 1.1 in Appendix 1.

2.2 Two stage

Referring to Fig A.2, and writing $C_1 R_1 = C_2 R_2 = T$ and $\omega T = x$, the voltage transfer function is given by $1/a_{11}$ of the resultant matrix.

$$\begin{bmatrix} 1 + jx & R \\ j\omega C & 1 \end{bmatrix}\begin{bmatrix} 1 + jx & R \\ j\omega C & 1 \end{bmatrix} = \begin{bmatrix} 1 + 3jx - x^2 & 2R + j\omega C R^2 \\ 2j\omega C - \omega^2 C^2 R & 1 + jx \end{bmatrix}$$

$$\therefore a_{11} = (1 - x^2) + 3jx$$

$$|a_{11}| = \sqrt{1 + x^4 + 7x^2} = \sqrt{2} \text{ at } f_0$$

$$\therefore x^2 = 0.14 \qquad \therefore \omega_0^2 = \frac{0.14}{T^2}$$

Relative response

$$\left|\frac{1}{a_{11}}\right| = \frac{1}{\sqrt{1 + 0.02(f/f_0)^4 + 0.98(f/f_0)^2}} \qquad \ldots [9]$$

Phase response

$$\phi = -\tan^{-1}\left[\frac{3\omega T}{1 - \omega^2 T^2}\right]$$

Relative phase response

$$\phi = -\tan^{-1}\left[\frac{1.12(f/f_0)}{1 - 0.14(f/f_0)^2}\right] \qquad \ldots [10]$$

2.3 Three stage

Manipulating the matrix product as above for three stages gives

$$a_{11} = (1 - 5x^2) + j(6x - x^3)$$

$$|a_{11}| = \sqrt{1 + x^6 + 13x^4 + 26x^2} = \sqrt{2} \text{ at } f_0$$

Solving gives $x^2 = 0.038$ $\qquad \therefore \omega_0^2 = \frac{0.038}{T^2}$

Relative response $= \dfrac{1}{\sqrt{1 + 5.4 \times 10^{-5}(f/f_0)^6 + 0.019(f/f_0)^4 + 0.981(f/f_0)^2}}$
$$\ldots [11]$$

Phase response

$$\phi = -\tan^{-1}\left[\frac{6\omega T - \omega^3 T^3}{1 - 5\omega^2 T^2}\right] \qquad \ldots [11a]$$

Relative phase response

$$\phi = -\tan^{-1}\left[\frac{1.17(f/f_0) - 0.0073(f/f_0)^3}{1 - 0.189(f/f_0)^2}\right] \qquad \ldots [12]$$

2.4 Four stage

$$a_{11} = (1 + x^4 - 15x^2) + j(10x - 7x^3)$$

$$|a_{11}| = \sqrt{1 + x^8 + 19x^6 + 87x^4 + 70x^2} = \sqrt{2} \text{ at } f_0$$

Solving gives $x^2 = 0.014$ $\qquad \therefore \omega_0^2 = \frac{0.014}{T^2}$

Relative response =

$$\frac{1}{\sqrt{1 + 3.9 \times 10^{-8}(f/f_0)^8 + 5.3 \times 10^{-5}(f/f_0)^6 + 0.017(f/f_0)^4 + 0.983(f/f_0)^2}}$$

... [13]

Phase response

$$\phi = -\tan^{-1}\left[\frac{10\omega T - 7\omega^3 T^3}{1 + \omega^4 T^4 - 15\omega^2 T^2}\right] \qquad ... [13a]$$

Relative phase response

$$\phi = -\tan^{-1}\left[\frac{1.185(f/f_0) - 0.012(f/f_0)^3}{1 + 1.97 \times 10^{-4}(f/f_0)^4 - 0.211(f/f_0)^2}\right] \quad ... [14]$$

Appendix 3. Methods for varying phase relationships without changing the amplitude

3.1 Passive circuit

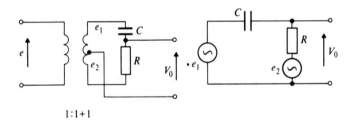

$$1:1+1$$

Fig. A.3

Figure A.3 shows one possible technique. It is limited to a.c. amplifiers due to the transformer coupling, but the transformer can be eliminated by the use of a phase-splitting amplifier arrangement to drive e_1 and e_2.

$$V_0 = \frac{R}{R + 1/j\omega C}e_1 + \frac{1/j\omega C}{R + 1/j\omega C}e_2$$

With a centre-tapped transformer, $-e_2 = +e_1 = e$

$$\therefore V_0 = e\frac{(-1 + j\omega CR)}{(1 + j\omega CR)} = \frac{-1 + \omega^2 C^2 R^2 + 2j\omega CR}{1 + \omega^2 C^2 R^2}e$$

$$|V_0| = e\sqrt{\frac{(\omega^2 C^2 R^2 - 1)^2 + 4\omega^2 C^2 R^2}{(1 + \omega^2 C^2 R^2)^2}}$$

$$= e\sqrt{\frac{(1 + \omega^2 C^2 R^2)^2}{(1 + \omega^2 C^2 R^2)^2}} = e \qquad ... [15]$$

The magnitude of V_o does not change with frequency. This is the ideal situation, of course, and a rigorous examination must include the reactive components of the non-ideal transformer. V_o also remains constant if R is changed, but as will be seen below, the phase shift can be adjusted.

$$\phi = \tan^{-1}\left[\frac{2\omega CR}{\omega^2 C^2 R^2 - 1}\right] \qquad \dots [16]$$

Let $\omega C = A$, then $\phi = \tan^{-1}\left[\dfrac{2AR}{A^2 R^2 - 1}\right]$

Let $u = \dfrac{2AR}{A^2 R^2 - 1}$

Then $\dfrac{\mathrm{d}\phi}{\mathrm{d}R} = \dfrac{\mathrm{d}\phi}{\mathrm{d}u}\dfrac{\mathrm{d}u}{\mathrm{d}R} = \dfrac{1}{1 + 4A^2 R^2/(A^2 R^2 - 1)^2}\dfrac{-2A(A^2 R^2 + 1)}{(A^2 R^2 - 1)^2}$

$$\frac{\mathrm{d}\phi}{\mathrm{d}R} = \frac{-2A(A^2 R^2 + 1)}{(A^2 R^2 - 1)^2 + 4A^2 R^2} = \frac{-2A}{A^2 R^2 - 1} \qquad \dots [17]$$

Thus ϕ is a non-linear function of R.

3.2 Active circuit

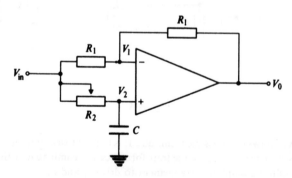

Fig. A.4

This can be used in direct-coupled applications. The inverting input terminal of Fig. A.4 receives the phase-shifted input signal, and the non-inverting input is forced to follow the same level (minus the error signal) as a result of closing the feedback loop.

Effective input signal $= V_2 - V_1$

$$V_{\text{eff}} = \frac{V_{\text{in}}}{1 + \mathrm{j}\omega CR_2} - \frac{V_{\text{in}}}{2}$$

G, closed-loop gain $\approx \dfrac{1}{\beta} \times V_{\text{eff}} = 2 \times V_{\text{eff}}$

$$= \frac{2}{1 + j\omega CR_2} - 1 = \frac{1 - j\omega CR_2}{1 + j\omega CR_2} = \frac{(1 - j\omega CR_2)^2}{1 + \omega^2 C^2 R_2{}^2}$$

$$|G| = \sqrt{\frac{(1 - \omega^2 C^2 R_2{}^2)^2 + 4\omega^2 C^2 R_2{}^2}{(1 + \omega^2 C^2 R_2{}^2)^2}} = 1 \qquad \dots [18]$$

and thus the magnitude of the gain is independent of the value of R_2 and the frequency.

Phase shift $\qquad\qquad \phi = -\tan^{-1}\left[\dfrac{2\omega CR_2}{1 - \omega^2 C^2 R_2{}^2}\right] \qquad\qquad \dots [19]$

Appendix 4. Common-mode errors resulting from capacitive mismatch

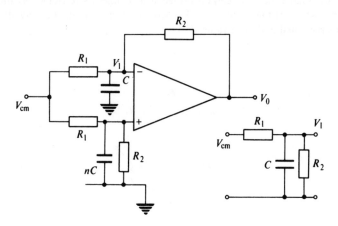

Fig. A.5

$$V_1 = V_{\text{cm}} \frac{R_2/(1 + j\omega CR_2)}{R_1 + R_2/(1 + j\omega CR_2)} = \frac{R_2}{(R_1 + R_2) + j\omega CR_1 R_2}$$

$$V_2 = V_{\text{cm}} \frac{R_2}{(R_1 + R_2) + jn\omega CR_1 R_2}$$

Let $a = R_1 + R_2$ and $b = \omega CR_1 R_2$, then the error signal $V_1 - V_2$

$$= V_{\text{cm}} R_2 \left(\frac{1}{a + jb} - \frac{1}{a + jnb}\right)$$

$$= V_{\text{cm}} R_2 \frac{jb(n-1)}{(a^2 - nb^2) + jab(1+n)} = \frac{ab^2(n-1)(n+1) + jb(n-1)(a^2 - nb^2)}{(a^2 - nb^2)^2 + a^2 b^2(1+n)^2} V_{\text{cm}} R_2$$

Writing magnitudes, $|V_1 - V_2|$

$$= V_{cm}R_2 \sqrt{\frac{a^2b^4(n-1)^2(n+1)^2 + b^2(n-1)^2(a^2-nb^2)^2}{[(a^2-nb^2)^2 + a^2b^2(1+n)^2]^2}}$$

$$= V_{cm}R_2 \frac{b(n-1)}{\sqrt{(a^2-nb^2)^2 + a^2b^2(1+n)^2}}$$

$$= V_{cm}R_2 \frac{\omega C R_1 R_2(n-1)}{\sqrt{[(R_1+R_2)^2 - n\omega^2C^2R_1{}^2R_2{}^2]^2 + (R_1+R_2)^2\omega^2C^2R_1{}^2R_2{}^2(1+n)^2}}$$

$$\ldots [20]$$

Appendix 5. Bridge transducer output

Different types of transducer tend to have a differing number of active arms; two- and four-arm bridges are most commonly used, but for completeness one and three active arms are considered here. Refer to Fig. A.6. In each case below, assume that the relevant bridge arm changes its resistance by δR in such a way that the output $V_2 - V_1$ increases in magnitude. Initially all the arms are equal and have value R.

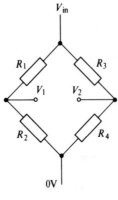

Fig. A.6

5.1 One active arm

$$R_1 = R + \delta R$$

$$\frac{V_1}{V_{in}} = \frac{R}{2R + \delta R}, \quad \frac{V_2}{V_{in}} = \frac{1}{2}$$

Output $V_1 - V_2 = V_o$

$$\frac{V_o}{V_{in}} = \frac{1}{2} - \frac{R}{2R+\delta R} = \frac{R}{4R+\delta R} = \frac{1}{4R/\delta R + 2} \approx \frac{\delta R}{4R} \quad \ldots [21]$$

The output of a bridge with a single active arm is fundamentally non-linear.

5.2 Two active arms

$$R_1 = R + \delta R, \quad R_2 = R - \delta R$$

$$\frac{V_o}{V_{in}} = \frac{1}{2} - \frac{R - \delta R}{2R} = \frac{\delta R}{2R} \qquad \qquad \ldots [22]$$

5.3 Three active arms

$$R_1 = R + \delta R \qquad R_2 = R - \delta R$$
$$R_3 = R - \delta R$$

$$\frac{V_1}{V_{in}} = \frac{R - \delta R}{2R} \qquad \frac{V_2}{V_{in}} = \frac{R}{2R - \delta R}$$

$$\therefore \frac{V_o}{V_{in}} = \frac{R}{2R - \delta R} - \frac{R - \delta R}{2R}$$

$$= \frac{3R\delta R - \delta R^2}{4R^2 - 2R\delta R} \qquad \qquad \ldots [23]$$

If δR is small, $\quad \dfrac{V_o}{V_{in}} = \dfrac{3R\delta R}{4R^2 - 2R\delta R} = \dfrac{3}{4R/\delta R - 2} \approx \dfrac{3\delta R}{4R}$

Again, a bridge with three active arms will be non-linear.

5.4 Four active arms

$\dfrac{V_o}{V_{in}}$ must be twice as large as that for a bridge with two active arms.

$$\text{i.e.} \quad \frac{V_o}{V_{in}} = \frac{\delta R}{R}$$

For a strain-gauge bridge with four active arms, $V_o = V_{in} \times$ gauge factor $\times$ strain, where gauge factor is defined as

$$\frac{\text{fractional change of resistance}}{\text{applied strain}} = \frac{\delta R/R}{S}$$

A convenient calibration technique for checking bridge integrity and amplifier gain is to shunt one bridge arm with a single fixed resistor. Refer to Fig A.7. For a bridge with a single active arm,

$$\frac{V_o}{V_{in}} \approx \frac{\delta R}{4R} = \frac{R - RR_x/(R + R_x)}{4R}$$

$$4R\frac{V_o}{V_{in}}(R + R_x) = R(R + R_x) - RR_x$$

$$\therefore 4\frac{V_o}{V_{in}}(R + R_x) = R$$

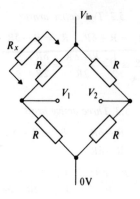

Fig. A.7

$$\therefore R_x = R\left(\frac{V_{\text{in}}}{4V_{\text{o}}} - 1\right) \qquad \ldots [24]$$

For example, $R = 350\,\Omega$, $V_{\text{in}} = 10\,\text{V}$, V_{o} required $= 10\,\text{mV}$

Then
$$R_x = 0.35\text{k}\left(\frac{10}{40 \times 10^{-3}} - 1\right)$$

$$= 87.2\,\text{k}\Omega$$

Appendix 6. 'Howland' current-output circuit

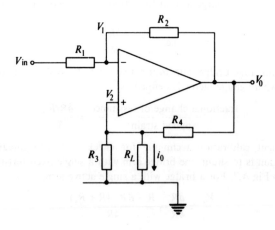

Fig. A.8

$$V_1 = \frac{R_2}{R_1 + R_2} V_{\text{in}} + \frac{R_1}{R_1 + R_2} V_{\text{o}} \qquad \ldots [25]$$

$$V_2 = \frac{R_L R_3 V_o}{R_4(R_L + R_3) + R_L R_3} \qquad \cdots [26]$$

$$V_2 = i_o R_L \qquad \cdots [27]$$

From [26] and [27],

$$V_o = \frac{i_o}{R_3}\left[R_4(R_L + R_3) + R_L R_3 \right] \qquad \cdots [28]$$

For high loop-gain situations, $V_1 = V_2$.

From [25] and [26],

$$V_{in}\frac{R_2}{R_1 + R_2} = V_o\left(\frac{R_L R_3}{R_L R_3 + R_4(R_L + R_3)} - \frac{R_1}{R_1 + R_2} \right) \qquad \cdots [29]$$

From [28] and [29],

$$i_o = V_{in}\frac{R_2}{R_1 + R_2}\frac{1}{R_L - \frac{R_1}{R_1 + R_2}\left(\frac{R_4}{R_3}(R_L + R_3) + R_L \right)}$$

Let
$$R_2 = \alpha R_1$$
$$R_4 = \alpha K R_1$$
$$R_3 = K R_1$$

$$i_o = \frac{\alpha}{1 + \alpha}V_{in}\frac{1}{R_L - \frac{1}{1 + \alpha}\left(\alpha(R_L + KR_1) + R_L \right)} \qquad \cdots [30]$$

$$i_o = \frac{-V_{in}}{KR_1} \qquad \cdots [31]$$

It follows from [31] that the output current is constant irrespective of the value of R_L.

Appendix 7. Constant current source

Assume that $i_1 \ll i_o$, so that $R_1 + R_2 \gg R_L$.

Then
$$V_1 = i_o R_L + V_Z$$
$$V_2 = i_o R_L + i_1 R_2$$
$$V_3 = i_o R_L + V_Z + i_o R_3$$
$$i_1 = \frac{V_3 - i_o R_L}{R_1 + R_2}$$

Combining these,

$$V_2 = i_o R_L + \frac{R_2}{R_1 + R_2}(V_3 - i_o R_L) = i_o R_L + \frac{R_2}{R_1 + R_2}(V_Z + i_o R_L)$$

$$\cdots [32]$$

and $V_2 = V_1$ if A is very large.

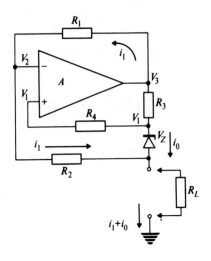

Fig. A.9

$$\therefore i_o R_L + \frac{R_2}{R_1 + R_2}(V_Z + i_o R_3) = i_o R_L + V_Z$$

$$\therefore i_o \frac{R_2 R_3}{R_1 + R_2} = V_Z\left(1 - \frac{R_2}{R_1 + R_2}\right) \qquad \dots [33]$$

$$\therefore i_o = V_Z\left(1 - \frac{R_2}{R_1 + R_2}\right)\frac{R_1 + R_2}{R_2 R_3} = V_Z\frac{R_1}{R_2 R_3} \qquad \dots [34]$$

If $$R_1 = R_2 \text{ then } i_o = \frac{V_Z}{R_3} \qquad \dots [35]$$

$$R_4 = \frac{R_1 R_2}{R_1 + R_2} \quad \text{for minimum drift.}$$

Appendix 8. Constant voltage across a remote variable load

Assume that R_3 and $R_4 \gg R_S$

$$i_o = \frac{V_5}{R_L} \qquad V_2 = \frac{R_2}{R_4 + R_2}V_4 \qquad V_4 = V_5\left(1 + \frac{R_C}{R_L}\right)$$

$$\therefore V_2 = \frac{R_2}{R_4 + R_2}V_5\left(1 + \frac{R_C}{R_L}\right) \qquad \dots [36]$$

$$V_1 = V_{in}\frac{R_3}{R_1 + R_3} + V_3\frac{R_1}{R_1 + R_3}, \qquad V_3 = V_4 + V_5\frac{R_S}{R_L}$$

$$\therefore V_1 = V_{in}\frac{R_3}{R_1 + R_3} + V_5\frac{R_1}{R_1 + R_3}\left(1 + \frac{R_S + R_C}{R_L}\right) \qquad \dots [37]$$

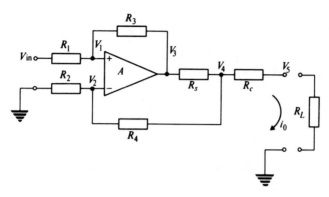

Fig. A.10

If A is very large, $V_1 = V_2$. Hence from [36] and [37],

$$V_{in}\frac{R_3}{R_1 + R_3} = V_5\left[\frac{R_2}{R_4 + R_2}\left(1 + \frac{R_C}{R_L}\right) - \frac{R_1}{R_1 + R_3}\left(1 + \frac{R_S + R_C}{R_L}\right)\right]$$

$$\dots [38]$$

For V_5 to be independent of R_L,

$$\frac{R_2}{R_4 + R_2}\frac{R_C}{R_L} = \frac{R_1}{R_1 + R_3}\frac{R_S + R_C}{R_L}$$

$$\therefore R_S = \left[\frac{R_2(R_1 + R_3)}{R_1(R_4 + R_2)} - 1\right]R_C \qquad \dots [39]$$

When this condition is fulfilled,

$$\frac{V_{in}}{V_5} = \frac{R_2(R_1 + R_3)}{R_3(R_4 + R_2)} - \frac{R_1}{R_3} \qquad \dots [40]$$

Example of component calculation

Consider a practical situation, where the line resistance $R_C = 10\ \Omega$. Choose $R_S = 8R_C = 80\ \Omega$

$$\text{From [39],} \quad \frac{R_2}{R_1}x = 9, \quad \text{where} \quad \frac{R_1 + R_3}{R_4 + R_2} = x$$

i.e. $$xR_2 = 9R_1 \qquad \dots [41]$$

From [40], $$\frac{V_{in}}{V_5} = x\frac{R_2}{R_3} - \frac{R_1}{R_3} = \frac{1}{R_3}(xR_2 - R_1) \qquad \dots [42]$$

From [41] and [42], $$\frac{V_{in}}{V_5} = \frac{8R_1}{R_3} \qquad \dots [43]$$

Let $\dfrac{V_{in}}{V_5} = 0.8$, then from [43], $R_3 = 10R_1$

Assuming a value of 10 kΩ for R_3, $R_1 = 1$ KΩ.

From [39], $\dfrac{11K}{1K}\dfrac{R_2}{R_4+R_2}=9$ $\therefore \dfrac{R_4}{R_2}=0.222$

Now if $R_2 = 10\,\text{k}\Omega$, $R_4 = 2.22\,\text{k}\Omega$

In this case, if $V_{\text{in}} = 4\,\text{V}$, $V_5 = 5\,\text{V}$

and
$$R_1 = 1\,\text{k}\Omega$$
$$R_2 = 10\,\text{k}\Omega$$
$$R_3 = 10\,\text{k}\Omega$$
$$R_4 = 2.22\,\text{k}\Omega$$

Appendix 9. Multi-path feedback

One possible network is that of Fig. 6.3(c), reproduced in Fig. A.11.

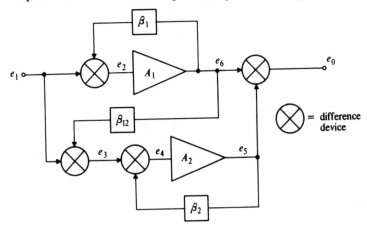

Fig. A.11

$$e_6 = \frac{A_1}{1+\beta_1 A_1}e_1$$

$$e_3 = e_1 - \beta_{12}e_6 = e_1\left(1 - \frac{\beta_{12}A_1}{1+\beta_1 A_1}\right)$$

$$e_5 = \frac{A_2}{1+\beta_2 A_2}e_3 = e_1\frac{A_2}{1+\beta_2 A_2}\left(1 - \frac{\beta_{12}A_1}{1+\beta_1 A_1}\right)$$

$$G = \frac{e_0}{e_1} = \frac{e_6 - e_5}{e_1} = \frac{A_1}{1+\beta_1 A_1} - \frac{A_2}{1+\beta_2 A_2}\left(1 - \frac{\beta_{12}A_1}{1+\beta_1 A_1}\right)$$
$$\dots [44]$$

Equation [44] gives the expression for overall gain. Stability of gain with variations of A_1 or A_2 can be obtained as follows.

$$\frac{dG}{G} = \frac{\partial G}{\partial A_1}\frac{dA_1}{G} + \frac{\partial G}{\partial A_2}\frac{dA_2}{G} \qquad \dots [45]$$

Taking each part of [45] separately,

$$\frac{\partial G}{\partial A_1} = \frac{1}{(1 + \beta_1 A_1)^2} \left(1 + \frac{\beta_{12} A_2}{1 + \beta_2 A_2} \right) \qquad \dots [46]$$

$$\frac{\partial G}{\partial A_2} = \frac{1}{(1 + \beta_2 A_2)^2} \left(\frac{\beta_{12} A_1}{1 + \beta_1 A_1} - 1 \right) \qquad \dots [47]$$

From [45],

$$\frac{dG}{G} = \frac{A_1}{G(1 + \beta_1 A_1)^2} \left(1 + \frac{\beta_{12} A_2}{1 + \beta_2 A_2} \right) \frac{dA_1}{A_1}$$

$$+ \frac{A_2}{G(1 + \beta_2 A_2)^2} \left(\frac{\beta_{12} A_1}{1 + \beta_1 A_1} - 1 \right) \frac{dA_2}{A_2} \qquad \dots [48]$$

Thus the fractional change of overall gain can be obtained from the fractional change dA_1/A_1 of A_1 and dA_2/A_2 of A_2. One method for the realisation of this circuit is shown in Fig. A.12.

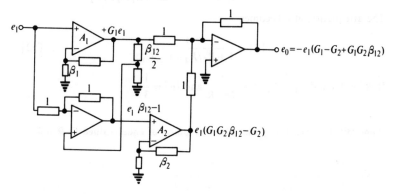

Fig. A.12

Appendix 10. Wien bridge network

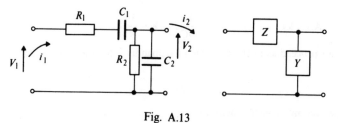

Fig. A.13

$$Z = R_1 + \frac{1}{j\omega C_1} \qquad Y = \frac{1}{R_2} + j\omega C_2$$

Writing the '*a*' matrix for the network,

$$\begin{bmatrix} V_1 \\ i_1 \end{bmatrix} = \begin{bmatrix} 1 & Z \\ 0 & 1 \end{bmatrix}\begin{bmatrix} 1 & 0 \\ Y & 1 \end{bmatrix}\begin{bmatrix} V_2 \\ -i_2 \end{bmatrix} \qquad \ldots [49]$$

$$= \begin{bmatrix} 1+ZY & Z \\ Y & 1 \end{bmatrix}\begin{bmatrix} V_2 \\ -i_2 \end{bmatrix}$$

The voltage transfer function is $F = \dfrac{1}{a_{11}} = \dfrac{1}{1+ZY}$

$$\therefore F = \frac{1}{(1+R_1/R_2+C_2/C_1)+\mathrm{j}[\omega C_2 R_1 - 1/(\omega C_1 R_2)]} \qquad \ldots [50]$$

The desired condition for use in an oscillator circuit is that the phase shift should be zero at a certain frequency. At f, then, the j-term must disappear.

$$\therefore \ \omega C_2 R_1 = \frac{1}{\omega C_1 R_2}, \qquad f_r = \frac{1}{2\pi \sqrt{R_1 R_2 C_1 C_2}} \qquad \ldots [51]$$

The attenuation at f_r becomes

$$|F| = \frac{1}{1+R_1/R_2+C_2/C_1} \qquad \ldots [52]$$

If $R_1 = R_2$ and $C_1 = C_2$, $f_r = \dfrac{1}{2\pi CR}$ and $|F| = \dfrac{1}{3}$.

Appendix 11. Frequency and phase response of equal-valued C–R ladder

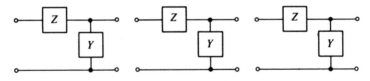

Fig. A.14

In Fig. A.14, let $\qquad Z = \dfrac{1}{\mathrm{j}\omega C}, \qquad Y = \dfrac{1}{R}$

We can proceed as in the example of Appendix 2 for the R–C ladder.

11.1

Single-section transfer function is

$$\frac{1}{\sqrt{1+1/x^2}} \qquad \ldots [53]$$

where $x = \omega CR$

Phase response $$\phi = \tan^{-1}\left(\frac{1}{x}\right) \qquad \ldots [54]$$

11.2

Two-section transfer function

$$F = \frac{1}{\sqrt{1 + 1/x^4 + 7/x^2}} \qquad \ldots [55]$$

Phase response $$\phi = \tan^{-1}\left[\frac{3x}{x^2 - 1}\right] \qquad \ldots [56]$$

11.3

Three-section $$F = \frac{1}{\sqrt{1 + 1/x^6 + 13/x^4 + 26/x^2}} \qquad \ldots [57]$$

$$\phi = \tan^{-1}\left[\frac{6x^2 - 1}{x^3 - 5x}\right] \qquad \ldots [58]$$

11.4

Four-section $$F = \frac{1}{\sqrt{1 + 1/x^8 + 19/x^6 + 87/x^4 + 70/x^2}} \qquad \ldots [59]$$

$$\phi = \tan^{-1}\left[\frac{10x^3 - 7x}{x^4 - 15x^2 + 1}\right] \qquad \ldots [60]$$

Appendix 12. Bridge-*T* network

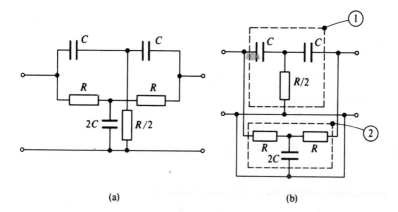

(a)   (b)

Fig. A.15

Figure A.15(a) can be redrawn as in Fig. A.15(b). One procedure for obtaining the transfer function using matrix techniques is as follows:

1. Obtain the 'a' matrix of each section.
2. Convert each to the equivalent 'y' matrix.
3. Add y_1 and y_2 to obtain the overall y.

(1) $Z = \dfrac{1}{j\omega C}$, $Y = \dfrac{2}{R}$

$$A_1 = \begin{bmatrix} 1 + \dfrac{2}{j\omega CR} & \dfrac{2}{j\omega C} - \dfrac{2}{\omega^2 C^2 R^2} \\[2mm] \dfrac{2}{R} & 1 + \dfrac{2}{j\omega CR} \end{bmatrix} \equiv \begin{bmatrix} \alpha & \beta \\ \gamma & \alpha \end{bmatrix} \qquad \dots [61]$$

(2) $Z = R$, $Y = 2j\omega C$

$$A_2 = \begin{bmatrix} 1 + 2j\omega CR & 2R + 2j\omega CR^2 \\ 2j\omega C & 1 + 2j\omega CR \end{bmatrix} \equiv \begin{bmatrix} \varepsilon & i \\ \kappa & \varepsilon \end{bmatrix} \qquad \dots [62]$$

Convert A_1 to y-parameters (using conversion tables),

$$A_1 = \begin{bmatrix} \dfrac{\alpha}{\beta} & -\dfrac{(\alpha^2 - \gamma\beta)}{\beta} \\[2mm] -\dfrac{1}{\beta} & \dfrac{\alpha}{\beta} \end{bmatrix}$$

but ΔA_1, the determinate $\dfrac{\alpha^2 - \gamma\beta}{\beta} = 1$

$$\therefore A_1 = y_1 = \begin{bmatrix} \dfrac{\alpha}{\beta} & -\dfrac{1}{\beta} \\[2mm] -\dfrac{1}{\beta} & \dfrac{\alpha}{\beta} \end{bmatrix} \qquad \dots [63]$$

Similarly,

$$y_2 = \begin{bmatrix} \dfrac{\varepsilon}{i} & -\dfrac{1}{i} \\[2mm] -\dfrac{1}{i} & \dfrac{\varepsilon}{i} \end{bmatrix} \qquad \dots [64]$$

The overall y-matrix becomes

$$y = \begin{bmatrix} \dfrac{\alpha}{\beta} + \dfrac{\varepsilon}{i} & -\left(\dfrac{1}{\beta} + \dfrac{1}{i} \right) \\[2mm] -\left(\dfrac{1}{\beta} + \dfrac{1}{i} \right) & \dfrac{\alpha}{\beta} + \dfrac{\varepsilon}{i} \end{bmatrix} \qquad \dots [65]$$

The transfer function for a y-parameter network is

$$F = -\frac{y_{21}}{y_{22}} = \frac{\beta + i}{i\alpha + \beta\varepsilon} \qquad \dots [66]$$

Substituting in [66] and simplifying gives

$$F = \frac{\omega CR - \omega^3 C^3 R^3 + j(\omega^2 C^2 R^2 - 1)}{5\omega CR - \omega^3 C^3 R^3 + j(5\omega^2 C^2 R^2 - 1)} \qquad \dots [67]$$

For infinite attenuation, $F = 0$, $\therefore\ \omega^3 C^3 R^3 = \omega CR$ and $\omega^2 C^2 R^2 = 1$

i.e. $f = \dfrac{1}{2\pi CR}$

Appendix 13. Sallen and Key active filter circuit

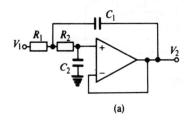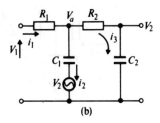

(a)   (b)

Fig. A.16

The lowpass circuit of Fig. A.16(a) can be redrawn as in Fig. A.16(b).

$$V_2 = \frac{i_3}{j\omega C_2} \qquad i_3 = j\omega C_2 V_2$$

$$V_a = V_2 + i_3 R_2 = V_2(1 + j\omega C_2 R_2) \qquad \dots [68]$$

$$V_1 = V_a + i_1 R_1 \qquad \dots [69]$$

$$i_1 = i_3 + i_2 = j\omega C_2 V_2 + i_2 \qquad \dots [70]$$

$$i_2 = (V_a - V_2)j\omega C_1 \qquad \dots [71]$$

From [68] and [71], $i_2 = -\omega^2 C_1 C_2 R_2 V_2$ $\qquad \dots [72]$

From [70] and [72], $i_1 = V_2(j\omega C_2 - \omega^2 C_1 C_2 R_2)$ $\qquad \dots [73]$

From [68], [69] and [73],

$$V_1 = V_2(1 + j\omega C_2 R_2) + V_2(j\omega C_2 R_1 - \omega^2 C_1 C_2 R_1 R_2)$$

Transfer function

$$\frac{V_2}{V_1} = \frac{1}{(1 - \omega^2 C_1 C_2 R_1 R_2) + j\omega(C_2 R_2 + C_2 R_1)}$$

$$= \frac{(1 - \omega^2 C_1 C_2 R_1 R_2) - j\omega(C_2 R_2 + C_2 R_1)}{(1 - \omega^2 C_1 C_2 R_1 R_2)^2 + \omega^2(C_2 R_2 + C_2 R_1)^2}$$

$$F = \frac{1}{\sqrt{(1 - \omega^2 C_1 R_1 C_2 R_2)^2 + \omega^2(C_2 R_2 + C_2 R_1)^2}} \qquad \dots [74]$$

Phase shift $\qquad \tan \phi = -\dfrac{(\omega C_2 R_2 + \omega C_2 R_1)}{1 - \omega^2 C_1 C_2 R_1 R_2}$ $\qquad$ ... [75]

If $C_1 R_1 = T_1$, $C_2 R_2 = T_2$, $C_2 R_1 = nT_1$, then

$$F = \frac{1}{\sqrt{(1 - \omega^2 T_1 T_2)^2 + \omega^2 (T_2 + nT_1)^2}}$$

Appendix 14. Capacitor-value modification

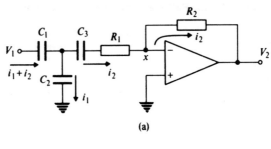

(a)

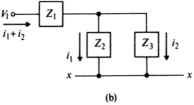

(b)

Fig. A.17

Figure A.17(b) represents the input circuit to the virtual earth point x. Then

$$Z_1 = \frac{1}{sC_1}, \quad Z_2 = \frac{1}{sC_2}, \quad Z_3 = \frac{1}{sC_3} + R_1$$

From circuit (b),

$$i_1 + i_2 = \frac{V_1}{Z_2 Z_3 / (Z_2 + Z_3) + Z_1} \qquad \ldots [76]$$

and $\qquad i_2 = \dfrac{Z_2}{Z_2 + Z_3}(i_1 + i_2) = \dfrac{Z_2 V_1}{Z_1 (Z_2 + Z_3) + Z_2 Z_3} \qquad \ldots [77]$

But i_2 is also equal to $-V_2 / R_2$ in circuit (a).

$\therefore$ from [77], $\qquad G = \dfrac{-V_2}{V_1} = \dfrac{-Z_2 R_2}{Z_1 (Z_2 + Z_3) + Z_2 Z_3}$

Substituting for the Z-values gives

$$G = \frac{-sC_1R_2}{(C_1+C_2)/C_3 + sR_1(C_1+C_2) + 1}$$

Appendix 15. Second-derivative circuit

The input network is represented by Fig. A.18(b) and the feedback network by Fig. A.18(c). In each case i_2 is identical.

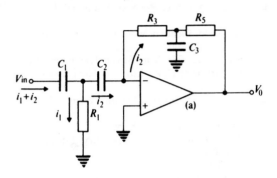

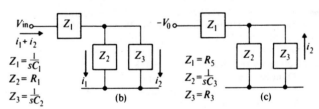

Fig. A.18

From circuit (b), $$i_2 = \frac{V_{in}Z_2}{Z_1(Z_2+Z_3)+Z_2Z_3}$$

$$= \frac{s^2C_1C_2R_1V_{in}}{1+sC_2R_1+sC_1R_1} \qquad \ldots [78]$$

From circuit (c), $$i_2 = \frac{-V_0}{R_5+sC_3R_3R_5+R_3} \qquad \ldots [79]$$

Equating [78] and [79] gives

$$V_0 = -V_{in} \times \frac{s^2C_1C_2R_3R_5(1/R_3+1/R_5+sC_3)}{1/R_1+s(C_1+C_2)} \qquad \ldots [80]$$

If $\dfrac{1}{R_3}+\dfrac{1}{R_5}+sC_3 = \dfrac{1}{R_1}+s(C_1+C_2),$

then $C_3 = C_1 + C_2$ and $R_1 = \dfrac{R_3 R_5}{R_3 + R_5}$

and $V_o = -V_{in} s^2 C_1 C_2 R_3 R_5 = -\dfrac{d^2 V_{in}}{dt^2} C_1 C_2 R_3 R_5$ ... [81]

Appendix 16. Principal manufacturers of linear integrated circuit operational amplifiers

Analog Devices Ltd, Central Avenue, East Molesey, Surrey.
Burr-Brown International Ltd, 17 Exchange Rd, Watford, Herts.
Datel UK Ltd, Stephenson Close, Portway Industrial Estate, Andover, Hants.
Delco Electronics AC-Delco, General Motors, Dunstable, Beds.
Exar Integrated Systems. Represented in UK by:
    Rastra Electronics Ltd, 275–281 King Street, Hammersmith, London.
Fairchild Camera & Instrument (UK) Ltd, 230 High Street, Potters Bar, Herts.
Ferranti Ltd, Electronic Components Division, Gem Mill, Chadderton, Oldham.
Harris Semiconductor. Represented in UK by:
    Memec Ltd, The Firs, Whitchurch, nr. Aylesbury, Bucks.
Hitachi. Represented in UK by:
    ITT Meridian, West Rd., Harlow, Essex.
Intersil Inc, 8 Tessa Rd, Richfield Trading Estate, Reading, Berks.
Lambda Electronics Ltd, Abbey Barn Rd, High Wycombe, Bucks.
Motorola Ltd, York House, Empire Way, Wembley, Middx.
Mullard Ltd, Mullard House, Torrington Place, London.
National Semiconductor UK Ltd, 19 Goldington Rd, Bedford.
Optical Electronics Inc. Represented in UK by:
    DI-AN Data Systems Ltd, 70–74 Princes Street, Stockport, Cheshire.
Plessey Semiconductors, Cheney Manor, Swindon, Wilts.
Precision Monolithics. Represented in UK by:
    Bourns (Trimpot) Ltd, Hodford House, 17/27 High St, Hounslow, Middx.
Raytheon Semiconductor, the Pinnacles, Harlow, Essex.
RCA Ltd Solid State-Europe, Sunbury-on-Thames, Middx.
Sescosem Thomson–CSF UK Ltd, Ringway House, Bell Rd, Daneshill, Basingstoke, Hants.
SGS–ATES (UK) Ltd, Planar House, Walton St, Aylesbury, Bucks.
Siemens Ltd, Siemens House, Windmill Rd, Sunbury on Thames, Middx.
Signetics – see Mullard Ltd
Siliconix Ltd, Morriston, Swansea.
Silicon General Inc. Represented in UK by:
    Rastra Electronics Ltd, 275–281 King St, Hammersmith, London.
Teledyne Philbrick, Cranford, Hounslow, Middx.
Texas Instruments Ltd, Manton Lane, Bedford.
Toshiba Semiconductors, Steatite Insulations Ltd, Hagley Rd, Birmingham

Appendix 17. Device recognition

Most manufacturers adopt a device coding system which makes devices recognisable and allows the engineer to determine the source without documentation. This can be useful for servicing and when replacements are required. Some prefixes are listed below to aid identification.

Type Number	Source
AD	Analog Devices
BB	Burr Brown
AM	Datel
DA	Delco Electronics
XR	Exar
μA, μAF	Fairchild
ZN	Ferranti
HA	Harris
ICL	Intersil
LAS	Lambda
MC, TCA	Motorola
TDA	Mullard
LM, LF, LH	National
OEI	Optical Electronics
SL, TAB	Plessey
SSS, OP-, PM	Precision Monolithics
RC, RM	Raytheon
CA	RCA
SF.C, TDA/B/C/etc.	Sescosem/Thomson−C.S.F.
TDA, L	SGS−ATES
TAA, TBA, TCA	Siemens
NE, SE, SA	Signetics
L	Siliconix
SG	Silicon General
ULN	Sprague
TP	Teledyne Philbrick
TLO, SN	Texas Instrument
TA	Toshiba
WC	Westinghouse

Appendix 18. Abbreviations

The following lists have been compiled to assist the student of electronics when confronted by the proliferation of terminology. The first part is confined to semiconductor devices both linear and digital, and the second part contains a few other abbreviations in common use which the circuit designer may require. With the rapid developments in electronic engineering, new terms are continuously emerging, so the following at best can be a mere guide to some of the more common expressions in use at the time of writing. There can be no substitute for the extensive reading of the literature required of any engineer in order to remain up to date.

18.1 Devices

AIM	Avalanche-induced migration
BBD	Bucket brigade device
BCL	Base coupled logic

BIFET	Bipolar output JFET input
BIMOS	Bipolar and MOS
CAM	Content addressable memory
CATT	Controlled avalanche transit time
CAZ	Commutating auto zero
C^3L	Complementary constant-current logic
CCD	Charge-coupled device
CDI	Collector diffusion isolation
CML	Current-mode logic
CMOS	Complementary MOS
CVD	Chemical vapour deposition
DI	Dielectric isolation
DIAC	Bi-directional a.c. rectifier
DIDO	Differential in, differential out
DIFMOS	Dual injector floating gate MOS
DIP	Dual in-line package
DMOS	Double-diffused MOS
DTL	Diode transistor logic
EAROM	Electrically alterable ROM
EBAM	Electron beam addressable memory
EBS	Electron bombarded semiconductor
ECL	Emitter coupled logic
EEPROM	Electrically erasable PROM
EFL	Emitter follower logic
EPROM	Erasable programmable ROM
ESFI	Epitaxial silicon film on insulator
FAMOS	Floating gate avalanche MOS
FET	Field-effect transistor
FIFO	First-in first-out register
FLOTOX	Floating gate tunnel oxide
FPLA	Field programmable logic array
FPROM	Field programmable ROM
GTOSCR	Gate turn-off SCR
HDCMOS	High-density CMOS
HINIL	High noise immunity logic
HLL	High-level logic
HSTTL	High-speed TTL
HTL	High-level transistor logic
IEC	Infused emitter coupling
IGFET	Insulated gate FET
I^2L	Integrated injection logic
I^3L	Isoplanar I^2L
JFET	Junction FET
JI	Junction isolation
LAPUT	Light-activated PUT
LASCR	Light-activated SCR
LCD	Liquid crystal display
LED	Light-emitting device (diode)
LIC	Linear integrated circuits
LIFO	Last in first out
LOCMOS	Locally oxidised CMOS

LPTTL	Low-power TTL
LSI	Large-scale integration
LS²	Low power Schottky (2)
MESFET	Metallised semiconductor FET
MIC	Microwave integrated circuit
MIS	Metal insulator silicon
MNOS	Metal nitride oxide semiconductor
MOS	Metal oxide semiconductor
MOSFET	Metal oxide semiconductor FET
MSI	Medium-scale integration
MTL	Merged transistor logic
NF	Noise figure
NMOS	N-channel MOS
PLA	Programmable logic array
PLL	Phase-locked loop
PMOS	P-channel MOS
PRAM	Programmable amplifier
PROM	Programmable read-only memory
PUT	Programmable UJT
RAM	Random access memory
ROM	Read-only memory
RPROM	RE-programmable PROM
RTL	Resistor transistor logic
SAW	Surface acoustic wave
SBS	Silicon bilateral switch
SCR	Silicon controlled rectifier
SCS	Silicon controlled switch
SFL	Substrate-fed logic
SIP	Single in-line package
SOAR	Safe operating area
SOS	Silicon on sapphire
STTL	Schottky TTL
TRIAC	Triggered bi-directional a.c. rectifier
TTL	Transistor-transistor logic
UART	Universal synchronous receiver/transmitter
UJT	Unijunction transistor
μPROC	Microprocessor
ULA	Uncommitted logic array
VI²L	V-Groove I²L
VLSI	Very large-scale integration
VMOST	Vertical channel MOS Transistor

18.2 Others

ADC	Analog–digital converter
ADP	Automatic data processing
AGC	Automatic gain control
ALU	Arithmetic logic unit
AM	Amplitude modulation
ANSII	American National Standards for information interchange
ASCII	American Standard Code for information interchange

ATE	Automatic test equipment
BCD	Binary coded decimal
BMS	Bit mark sequence
CODEC	Coder/decoder
CNRZ	Complementary NRZ
CPU	Central processing unit
CVSD	Continuously variable slope delta modulator
CW	Continuous wave
DAC	Digital–analog converter
dmm	Digital multimeter
D/S	Digital–synchro converter
DVM	Digital voltmeter
ECC	Error-correction code
ECMA	European computer manufacturers' association
EDP	Electronic data processing
FDNR	Functionally dependent negative resistor
FFT	Fast Fourier transform
FM	Frequency modulation
FSK	Frequency shift keying
FVC	Frequency-to-voltage converter
HF	High frequency
I/O	Input/Output (port)
LCD	Liquid crystal display
LF	Low frequency
MBM	Magnetic bubble memory
MODEM	Modulation–demodulation (equipment)
MPU	Microprocessor unit
MTBF	Mean time between failures
MTTF	Mean time to failure
MUX	Multiplexer
NRZ	Non-return to zero
PCB	Printed circuit board
PCM	Pulse-code modulation
pdf	Probability distribution function
PMUX	Programmable multiplexer
ppm	Parts per million
PPM	Pulse-position modulation
PRF	Pulse repetition frequency
PWM	Pulse-width modulation
RFI	Radio frequency interference
RTD	Resistance temperature detector
RZ	Return to zero
SBC	Single-board computer
S/D	Synchro–digital converter
SHF	Super high frequency
SMPS	Switched-mode power supply
SSB	Single sideband
TRF	Tuned radio frequency
TWT	Travelling wave tube
UHF	Ultra high frequency
VDU	Visual display unit

VF	Video frequency, voice frequency
VFC	Voltage-to-frequency converter
VHF	Very high frequency
VLF	Very low frequency
VOM	Volt-ohm-meter
VSWR	Vertical standing wave ratio

Appendix 19 dB attenuation vs. percentage error

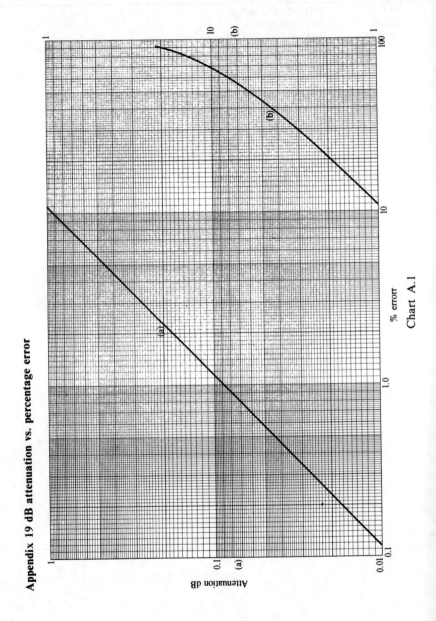

Chart A.1

References used in
the text

1 G. W. A. Dummer, 'Electronic components – Past present and future', *Electronic Components*, October 1970. U.T.P.
2 S. Givens, 'Precision I.C. op amps – underspecifying can mean disaster, overspecifying costs money', *E.D.N.*, 5 Sept. 1977.
3 J. Millman and H. Taub, *Pulse and Digital Circuits*, McGraw-Hill, 1956.
4 J. L. Marshall, *Lightning Protection*, Wiley Inter-science, 1973.
5 G. B. Clayton, *Operational Amplifiers*, Newnes-Butterworth, 1979.
6 H. W. Bode, *Network Analysis and Feedback Amplifier Design*, D. Van Nostrand, 1945.
7 D. E. Thomas, 'Tables of phase associated with a semi-infinite unit slope of attenuation', *Bell System Tech. J.*, 1947.
8 R. W. J. Barker, 'Compensation techniques for operational amplifiers', *Electronic Components*, 23 July 1971.
9 Karl Huehne, 'Using transient response to determine operational amplifier stability', *Motorola Application Note AN-460*, 1969.
10 Ralph Morrison, *Grounding and Shielding Techniques in Instrumentation*, Wiley, 1977.
11 P. Shillito, 'Slew rate limiting: Design nomogram', *Electron*, 24 Oct. 1974.
12 J. G. Graeme, *Applications of Operational Amplifiers*, McGraw-Hill, 1973 and G. E. Tobey, J. G. Graeme, and L. P. Huelsman, *Operational Amplifiers*, McGraw-Hill, 1971.
13 R. D. Middlebrook, *Differential Amplifiers*, Wiley, 1963.
14 R. I. Demrow, 'Narrowing the margin of error', *Electronics*, 15 April 1968.
15 Donn Soderquist, 'Minimisation of noise in operational amplifier applications', *Precision Monolithics Inc*, AN-15, 1975.
16 L. Smith and D. Sheingold, 'Noise and operational amplifier circuits', *Analog Dialogue*, 3, 1 March 1969.
17 T. J. Robe, 'Measurement of burst (popcorn) noise in linear integrated circuits', *R.C.A. Application Note ICAN-6732*, 1971.
18 C. N. G. Matthews, 'Noise figure measurement', *Wireless World*, August 1967.
19 Bernard Cole, 'IC simulates matched transistor pairs', *Electronics*, 28 Oct. 1976.

20 B. Sundquist, 'Transient intermodulation in amplifiers', *Wireless World*, Feb. 1977.

21 E. A. Goldberg, 'Stabilization of wide band direct current amplifiers for zero and gain', *R.C.A. Review*, June 1950.

22 I. C. Hutcheon and D. Summers, 'A low drift transistor chopper type d.c. amplifier with high gain and large dynamic range', *Proc. IEE*, March 1960.

23 I. C. Hutcheon, 'Properties of some d.c.–a.c. chopper circuits', *Proc. IEE*, Jan. 1957.

24 J. Watson, *An Introduction to Field Effect Transistors*, Siliconix Ltd, 1970.

25 T. K. Hemingway, *Electronic Designer's Handbook*, Business Publications Ltd, 1966.

26 W. Gosling, W. Townsend and J. Watson, *Field Effect Electronics*, Butterworth, 1971.

27 M. Stevens and A. Wood, 'The FET as a chopper', Texas Instruments Ltd, *Application Report B60*, 1968.

28 I. E. Shepherd, 'Sources of lost gain in chopper amplifiers', *Electron*, 28 Dec. 1972.

29 S. K. Bhola and R. H. Murphy, 'Design of low level d.c. chopper amplifiers', *Electronic Engineering*, May 1967.

30 P. Zicko, 'Designing with chopper stabilised operational amplifiers', *Design Electronics*, Sept. 1970.

31 R. C. Jaeger and G. A. Hellworth, 'Dynamic zero correction method suppresses offset error in op amps', *Electronics*, 4 Dec. 1972.

32 R. J. Widlar, 'Drift compensation techniques for integrated d.c. amplifiers', *National Semiconductor Corp. AN-3*, 1967.

33 R. J. Widlar, 'Linear ICs: Part 6. Compensating for drift', *Electronics*, 5 Feb. 1968.

34 A. G. J. MacFarlane, 'Feedback', *Measurement and Control*, 9, 12, Dec. 1976.

35 H. S. Black, 'Translating system', US Patent 1686792, 9 Oct. 1928; filed 3 Feb. 1925.

36 H. W. Bode, 'Feedback – The history of an idea', *Symposium on Active Networks and Feedback Systems*, Polytechnic Institute of Brooklyn, April 1960.

37 H. S. Black, 'Wave translating system', US Patent 2102671, 21 Dec. 1937.

38 'Improvements in or relating to arrangements for amplifying electrical oscillation', British Patent 323823, 16 Jan. 1930.

39 C. J. Van Loon, 'Improvements in radio receivers', *Philips Tech. Rev.*, 1, 9, Sept. 1936.

40 W. Six and H. Mulders, 'The use of amplifiers (repeaters) in telephony', *Philips Tech. Rev.*, 2, 7, July 1937.

41 B. D. H. Tellegen and V. Cohen Henriquez, 'Inverse feed-back', *The Wireless Engineer*, August 1937.

42 B. D. H. Tellegen, 'Inverse feed-back', *Philips Tech. Rev.*, 2, 10, October 1937.

43 'Feedforward compensation speeds op amp', National Semiconductor Corp. *Linear Brief*, 2, 1969.

44 E. Renschler, 'The MC 1539 operational amplifier and its applications', *Motorola Applications Note AN-439*, 1972.
45 J. J. Zaalberg van Zelst, 'Stabilised amplifiers', *Philips Tech. Rev.*, 9, 1, 1947.
46 J. J. Zaalberg van Zelst, 'Constant amplification in spite of changeability of the circuit elements' *Philips Tech. Rev.*, 9, 10, 1947.
47 W. D. Lewis, 'Self correcting amplifier', US Patent 2592716, 15 April 1952.
48 H. Seidel, H. R. Beurrier and A. N. Friedman, 'Error controlled high power linear amplifiers at V.H.F.', *Bell System Tech. J.*, May–June 1968.
49 H. Seidel, 'A feedforward experiment applied to an L-4 carrier system amplifier' *IEEE Trans. on Comm. Tech.*, Vol. Comm-19, 3, June 1971.
50 H. Seidel, 'A microwave feedforward experiment', *Bell System Tech. J.*, 50, 9, Nov. 1971.
51 'Feedforward error control', *Wireless World*, May 1972.
52 R. K. Jurgen, 'Feedforward correction – a late blooming design', *IEEE Spectrum*, April 1972.
53 T. J. Bennett and R. F. Clements, 'Feedforward – an alternative approach to amplifier linearisation', *The Radio and Electronic Engineer*, 44, 5, May 1974.
54 J. J. Golombeski, M. S. Ghausi *et al.*, 'A class of minimum sensitivity amplifiers', *IEEE Trans. on Circuit Theory*, 14, 1, March 1967.
55 E. H. Cooke Yarborough, M. O. Deighton and C. L. Miller, 'A method of enhancing the transient response stability of fast transistor amplifier systems', *UKAEA Research Group Report AERE-R6851*, 1971.
56 B. McMillan, 'Multiple feedback systems', US Patent 2748201, 29 May 1956.
57 A. M. Sandman, 'Reducing amplifier distortion', *Wireless World*, Oct. 1974.
58 P. J. Walker, 'Current dumping audio amplifier', *Wireless World*, Dec. 1975.
59 E. M. Cherry and D. E. Hooper, *Amplifying Devices and Low Pass Amplifier Designs*, Wiley, 1968.
60 G. J. Pridham, 'Analysis of feedback amplifiers', *Electronic Engineering*, July 1967.
61 D. Sheingold, 'Simple rules for choosing resistor values in adder–subtractor circuits', *Analog Dialogue*, 10, 1, 1976.
62 D. K. Fryer, 'Maintaining a constant voltage across a remote variable load', *New Electronics*, 8 March 1977.
63 H. W. Bode, 'Relations between attenuation and phase in feedback amplifier design', *Bell System Tech. J.*; also *Network Analysis and Feedback Amplifier Design*, Van Nostrand, 1945.
64 Thomas Roddam, 'The Bode fillet', *Wireless World*, Feb. 1964.
65 Jefferson, *Proc. Inst. Radio Engineers*, 39 (1951), p. 1571.
66 Edwin and Roddam, *Principles of Feedback Design*, Iliffe, 1964.
67 Felker 'Calculator and chart for feedback problems' *Proc. Inst. Radio Engineers*, 37, 1204, 1949.
68 H. Nyquist, 'Regeneration theory', *Bell System Tech. J.*, 11 Jan. 1932.
69 Cushman, 'Stability and relative stability' in Truxal, (ed.), *Control Engineer's Handbook*, McGraw-Hill.

70 J. J. Di Stefano, A. R. Stubberud, and I. J. Williams, *Feedback and Control Systems*, McGraw-Hill, 1967.

71 D. S. Cutler 'Closed loop response of a control system', *New Electronics*, 4 Oct. 1977.

72 'Linear applications', National Semiconductor Corp.

73 'Linear integrated circuits', R.C.A.

74 'Linear application notes', Motorola Semiconductor Corp.

75 J. G. Graeme, *Designing with Operational Amplifiers*, McGraw-Hill for Burr-Brown, 1977.

76 Y. J. Wang and W. E. Ott, *Function Circuits*, McGraw-Hill for Burr-Brown, 1976.

77 G. Bajen, 'Radio frequency LC oscillators', *Electronic Engineering*, Oct. 1977.

78 R. N. Caffin, 'Simple low distortion Wien bridge oscillator', *Electronic Engineering*, Oct. 1975.

79 D. Hileman, 'Common silicon diodes stabilize oscillator', *Electronics*, 9 Jan. 1975.

80 V. B. Mehta, 'Distortion in RC bridge feedback oscillators', *Electronic Engineering*, Sept. 1967, pp. 582–5.

81 K. G. Beauchamp, 'A twin-T filter design having an adjustable centre frequency', *Electronic Engineering*, June 1967.

82 F. Butler, 'Transistor R-C oscillators and selective amplifiers', *Wireless World*, Dec. 1962.

83 R. P. Sallen and E. L. Key, 'A practical method of designing RC active filters', *IRE Trans on Circuit Theory*, March 1955.

84 M. Baril, 'Three mode network makes filter or oscillator', *Electronics*, 12 April 1973.

85 W. G. Jung, 'Low distortion oscillator', *Electronics*, 5 Feb. 1976.

86 Y. J. Wong and W. E. Ott, *Function Circuits*, McGraw-Hill for Burr-Brown, 1976.

87 B. E. Jones, *Instrumentation, Measurement and Feedback*, McGraw-Hill, 1977.

88 H. K. P. Neubert, *Instrument Transducers*, O.U.P., 1975.

89 H. K. P. Neubert, *Strain Gauges – Kinds and Uses*, Macmillan, 1967.

90 R. J. Redding, *Intrinsic Safety*, McGraw-Hill, 1971.

91 D. Weatherhead, 'Intrinsic safety', *Measurement and Control*, **10**, Sept. 1977.

92 'Intrinsic Safety', BASEEFA Buxton, Derbyshire, SFA 3012.

93 BS5345, Part 4.

94 B. Stevens, 'Signal transmission put on a pedestal', *Control and Instrumentation*, Sept. 1976.

95 R. K. Demrow, 'Narrowing the margin of error', *Electronics*, 15 April 1968.

96 I. E. Shepherd, 'Some properties of wire resistance strain gauges', *Electron*, 17 July 1975.

97 M. A. Weiner, 'Quadrature filters: What they are and what they do', *The Electronic Engineer*, Sept. 1971.

98 J. C. S. Richards, 'Capacitor fed parallel chopper as a phase sensitive demodulator', *Proc. IEE*, **118**, 12, Dec. 1971.

99 N. Pollock, 'A simple high performance device for measuring strain gauge transducer outputs', *J. phys. E: Scientific Instruments*, **8**, 1975.

100 M. Kreuzer, *Design Concept of a New High Resolution Measuring Device and its Application in the Fields of Electromechanical Weighing and Force Measurement*, Hottinger Baldwin Mestechnik GmbH.

101 H. Krabbe, 'Monolithic data amplifier', *Analog Dialogue*, 6, 1, Spring 1972.

102 G. J. Clarke, 'Linear isolators for clean, safe, signal transmission', *Control and Instrumentation*, Nov. 1977.

103 T. Gartner, 'IC timer and voltage doubler form a dc–dc converter', *Electronics*, 22 August 1974.

104 C. Scott and R. Stitt, 'Inverting dc–dc converters require no inductors', *Electronics*, 22 Jan. 1976.

105 I. E. Shepherd 'Low volt-drop regulator', *Electronic Engineering*, Jan. 1973.

106 I. E. Shepherd, 'Three terminal micropower voltage regulator', *Electronic Engineering*, Oct. 1977.

107 I. E. Shepherd, 'Transmitter has low operating voltage', *Control and Instrumentation*, Feb. 1973.

108 R. J. Widlar, R. C. Dobkin and M. Yamatake, 'New op amp ideas', *National Semiconductor Corp. AN 211*, 1978.

109 *Piezoelectric Accelerometer User's Handbook*, Bell & Howell Ltd.

110 R. K. Underwood, 'New design techniques for FET op amps', *National Semiconductor Corp. AN-63*, 1972.

111 R. Bird, 'Obtaining a voltage–time derivative from slowly varying voltages', *J. Phys. E: Scientific Instruments*, 10, 10, Oct. 1977.

112 R. Stata, 'Operational integrators', *Analog Dialogue*, 1, 1, April 1967.

113 'Switching mode power supplies', Texas Instruments Ltd, *Applications Report B76*. Also, 'Industrial switching mode power supplies', *B159*.

114 R. J. Widlar, 'Designing switching regulators', *National Semiconductor Corp. AN-2*, 1969.

115 'Designs for negative voltage regulators', *National Semiconductor Corp. AN-21*, 1968.

116 T. M. Frederiksen, W. M. Howard and R. S. Sleeth, 'The LM3900 – A new current differencing quad of ±input amplifiers', *National Semiconductor Corp. AN-72*, 1972.

117 'Applications of the CA3080 high performance operational transconductance amplifiers', RCA Corp., *ICAN6668*.

118 R. Wagner, 'Laser trimming on the wafer', *Analog Dialogue*, 9, 3, 1975.

119 J. Metzger, 'Bipolar and FET op amps', *Electronic Products*, June 1977.

120 R. Russell, 'BI–FET: A new linear integrated circuit', *Electron*, Jan. 1978.

121 R. J. Widlar, 'New process makes possible 1 Volt linears', *Electronics*, 2 March 1978.

122 R. J. Widlar, R. C. Dobkin and M. Yamatake, 'Low Voltage Op Amp breakthrough expands linear design horizons. *E.D.N.*, 5 Feb. 1979.

Bibliography

1   Hnatek, *Application of Linear Integrated Circuits*, Wiley, 1975.
2   B. E. Jones, *Instrumentation, Measurement and Feedback*, McGraw-Hill, 1977.
3   Coughlin and Driscoll, *Operational Amplifiers and Linear Integrated Circuits* Prentice-Hall, 1977.
4   Connelly (ed.), *Analogue Integrated Circuits*, Wiley, 1975.
5   Bannon, *Operational Amplifiers – Theory and Servicing*, Reston, 1975.
6   G. D. Bishop, *Linear Electronic Circuits and Systems*, Macmillan, 1974.
7   J. D. Lenk, *Operational Amplifier Users (Manual for)*, Reston, 1976.
8   Roberge, *Operational Amplifiers – Theory and Practice*, Wiley, 1975.
9   Ratkowski, *Handbook of Integrated Circuit Operational Amplifier*, Prentice-Hall, 1975.
10 D. F. Stout and M. Kaufman, *Handbook of Operational Amplifier Circuit Design*, McGraw-Hill, 1976.
11 Kalvoda, *Operational Amplifiers in Chemical Instrumentation*, Ellis Horwood, 1975.
12 J. G. Graeme, *Designing with Operational Amplifiers – Applications Alternatives*, McGraw-Hill, 1977.
13 Y. J. Wong and W. E. Ott, *Function Circuits*, McGraw-Hill, 1976.
14 J. B. Eimbinder (ed.), *Designing with Linear Integrated Circuits*, Wiley, 1969.
15 J. B. Eimbinder (ed.), *Linear Integrated Circuits – Theory and Applications*, Wiley, 1969.
16 T. D. S. Hamilton, *Handbook of Linear Integrated Electronics for Research*, McGraw-Hill, 1977.
17 G. B. Clayton, *Linear Integrated Circuit Applications*, Macmillan, 1975.
18 G. B. Clayton, *Experiments with Operational Amplifiers*, Macmillan, 1975.
19 J. B. Dance, *Op-Ampo*, Newnes, 1978.
20 J. J. Carr, *Op Amp Circuit Design and Applications*, Foulsham-Tab, 1976.
21 J. V. Wait, L. P. Huelsman and G. A. Korn, *Introduction to Operational Amplifier Theory and Applications*, McGraw-Hill – Kogakusha, 1975.
22 G. B. Clayton, *Operational Amplifiers*, Newnes Butterworth, 1979.
23 H. M. Berlin, *Design of Operational Amplifier Circuits,* Sams, 1977.

Index